MOON RUSH

ALSO BY LEONARD DAVID

(with Buzz Aldrin) *Mission to Mars: My Vision for Space Exploration*

Mars: Our Future on the Red Planet

MOON RUSH

THE NEW SPACE RACE

LEONARD DAVID

WASHINGTON, DC

Published by National Geographic Partners, LLC
1145 17th Street NW Washington, DC 20036

ISBN: 978-1-4262-2005-0

Since 1888, the National Geographic Society has funded more than 13,000 research, exploration, and preservation projects around the world. National Geographic Partners distributes a portion of the funds it receives from your purchase to National Geographic Society to support programs including the conservation of animals and their habitats.

Get closer to National Geographic explorers and photographers, and connect with our global community. Join us today at nationalgeographic.com/join
For information about special discounts for bulk purchases, please contact National Geographic Books Special Sales: specialsales@natgeo.com

For rights or permissions inquiries, please contact National Geographic Books Subsidiary Rights: bookrights@natgeo.com

Printed in the United States of America

Dedicated to Paul Spudis and Larry Taylor

The loss of these two giants of lunar science
and close personal friends of the author
occurred during the writing of this book.

THE ORIGINAL MOON

BY CARLE PIETERS

Four and a half aeons ago
a dark, dusty cloud deformed.
Sun became star; Earth became large,
and Moon, a new world, was born.

This Earth/Moon pair, once linked so close,
would later be forced apart.
Images of young intimate ties,
we only perceive in part

Both Earth and Moon were strongly stripped
of their mantle siderophiles,
But Moon alone was doomed to thirst
from depletion of volatiles.

Moon holds secrets of ages past
when planets dueled for space.
As primordial crust evolved
raw violence reworked Moon's face.

After the first half billion years
huge permanent scars appeared.
Ancient feldspathic crust survived
with a mafic mantle mirror.

But then there grew from half-lived depths
a new warmth set free inside.
Rivers and floods of partial melt
resurfaced the low "frontside."

Thus evolved the Original Moon
in those turbulent times.
Now we paint from fragments of clues
the reasons and the rhymes:

Sister planet;
Modified clone;
Captured migrant;
Big Splash disowned?

The Truth in some or all of these
will tickle, delight,
temper, and tease.

CONTENTS

FOREWORD

LET'S GET OUR GLOBAL SPACE ACT TOGETHER

BY BUZZ ALDRIN

LUNAR MODULE PILOT, APOLLO 11

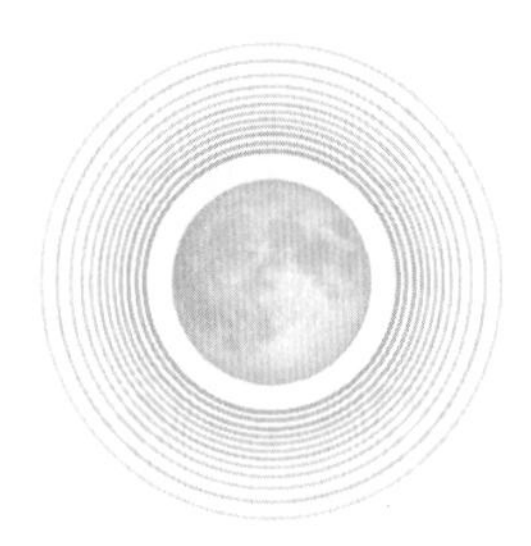

FIFTY YEARS AGO, ON JULY 20, 1969, I know exactly where I was standing. I was well suited for where I was, but without a tie. My get-down-to-business attire weighed 180 pounds. After stepping down from the *Eagle* lander, I stood there with my fellow space traveler, Neil Armstrong. High above the barren lunar scenery that I viewed as "magnificent desolation" was our Apollo 11 partner, Mike Collins, orbiting the Moon.

America's Apollo program was an impressive team effort. It drew upon the talents of thousands and thousands of individuals who came together to make real a vision. It was a unified undertaking: a mix of government, industry, and academic stamina to transform a long-held dream into a reality.

But let's focus on a future trajectory. We need to revitalize the U.S. space program and fortify a leadership position, being mindful of the growing stature of other spacefaring nations that also are setting their sights on the heavens. I am more optimistic today than just a few years ago, given the revitalization of the National Space Council, under the chairmanship of Vice President Mike Pence. This council can guide American space policy in the direction I assert we should be moving: to "regain and retain" U.S. leadership in space. Our nation has suffered by not having an energetic, resolute, and true trajectory for NASA's exploration program.

That given, here's my vision for the future.

From a U.S. leadership position, we need to pull together space-capable nations to forge a partnership. I call it a global Moon-Mars coalition of nations, including China. The United States should stand forth as the experienced and essential leader of this coalition of countries determined to access space. By forming this international coalition, we will decrease for all the cost of many activities in Earth orbit and at the Moon in preparation for our missions to Mars.

But let me be blunt about something.

I have a major message to ground control—and, for the most part, ground control is at NASA headquarters in Washington, D.C. My communiqué consists of two words: "fiscal responsibility." Also, as in Apollo, an authoritative, independent advisory group should be available to those at NASA headquarters and others who are blueprinting the future with the intent of furthering a global space coalition. Space control should be in Houston, Texas—a mission control to be independent from politics. There are major decisions to be made—whether about the utility of what's dubbed the Lunar Orbital Platform-Gateway that's now in vogue, or how long to keep the International Space Station (ISS) on a financial lifeline, or the proper utilization and future of NASA's Orion spacecraft, and also NASA's new big booster, the Space Launch System. Great stewardship of these programs is necessary; otherwise we will eat up every piece of the NASA budget, and we won't get anywhere.

As we celebrate the 50th anniversary of the landing of Apollo 11's *Eagle*—the spacecraft that got Neil and me down to the Moon's surface safe and sound—let's rekindle that momentum of progress. I see any number of *Eagle* landers in the future. There's a clear opportunity to do that.

I am proud that I have championed the call for establishing a permanent human settlement on Mars over the decades. Perhaps, on this historic occasion of saluting the first human landing on the Moon, we need to signal future goals, those that inspire and draw upon the wherewithal of the global space community to push forward, not only to Mars, but beyond.

There is a historical legacy, and we humans need to live up to that heritage. Our Moon, as you can read throughout this book, is a world in waiting. How we utilize Earth's celestial neighbor to further space objectives is yet to be determined. What I do know is that beyond returning to the Moon, occupying Mars is a task like no other. That undertaking, I firmly believe, can unite the great nations of the world in a cooperative, beneficial way. We can set sail to Mars, eventually putting in place a flourishing civilization step-by-step. This peaceful pursuit is unparalleled in history. It is time to place spacefaring nations on that unified trajectory.

I pride myself in supporting farsighted goals and objectives, a mandatory trait of space exploration planning, but I never forget that failure is an option. Reminder: Don't get bogged down wearing 180 pounds of Moon suit! Let's roll up our sleeves and make the future happen.

CHAPTER ONE

THE MOON: WHAT'S THE ATTRACTION?

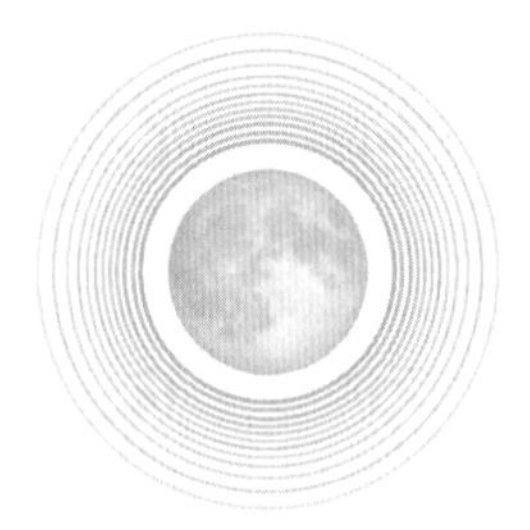

IN THE MID-1950S, it was easy to fly to the Moon. All you needed was a ticket to Tomorrowland. Walt Disney himself called this portion of Disneyland "a vista into a world of wondrous ideas, signifying man's achievements" and "a step into the future, with predictions of constructive things to come." There it stood, a stunning eyeful, a red-and-white 72-foot Rocket to the Moon.

I remember it well. Rocket travelers, including myself, sat in seats encircling a screen on the floor. Above us, another circular screen was mounted on the ceiling. At liftoff, the seats vibrated, even inflated (to portray weightlessness, I guess), as we space travelers headed moonward. The trek took 10 thrilling minutes as we watched the lunar panorama whiz by. A disembodied voice of wisdom explained the orb we were approaching. As I recall from my youth, we swung around the far side of the Moon, and flares lit up the Moon's dark terrain, revealing some kind of structure—perhaps a city from ancient times? Was it an advanced civilization hidden from us until now? No time to ponder—strapped into our seats, we were on our way back home to a safe touchdown at the California spaceport, earning a certificate signed by the vessel's steely-nerved captain, verifying that we had rocketed around the Moon.

A lot has happened since then. First sponsored by Trans World Airlines (TWA), then Douglas Aircraft, and finally the aerospace firm McDonnell Douglas, Rocket to the Moon was replaced by Flight to the Moon in 1967, then discontinued in 1975 and replaced by Mission to Mars. The original Disneyland spaceport was replaced by a Pizza Port. A sign of the times? Corporate takeovers, shifting missions, commercial ventures edging out research and exploration?

It's time to go back to the Moon, and not just an imaginary Disneyland adventure. The Moon is a footprint-friendly place, as we know, thanks to the extraordinary human missions of the late 1960s and early 1970s. The 50-year legacy of Project Apollo is a foundation for the future—but what will that future look like?

—||—

THE ONE HEAVENLY OBJECT easily seen by all on Earth, the Moon has intrigued humankind throughout history. Lunar deities appear in just about every mythological tradition. Thoth, an Egyptian god of writing, magic, and wisdom, was associated with the Moon's mysteries. Egyptians also worshipped the Moon goddess Isis as the Star of the Sea. The oceanic Sina, Polynesian deity of love, beauty, and fertility, lives in the Moon itself, guarding those who travel at night. In ancient Roman religion and myth, the goddess Luna is the divine embodiment of the Moon; her Greek counterpart was Selene, the goddess who drives her Moon chariot through the heavens. The Hindu deity was Chandra, from the word for "shining" or "moon"; the Chinese goddess Chang'e lives in the Moon with the Jade Rabbit. Both names, Chandra and Chang'e, are back in the news as mission monikers. The Moon has played a major part in Native American life; among the Hopi in Northern Arizona, for example, observations of the Moon set the time of women's ceremonies.

Hundreds of songs have lunar lyrics. "Moon River," "Moonshadow," "Bad Moon Rising," "It's Only a Paper Moon"—the list goes on and on, so many songwriters and crooners have used the Moon as a theme to showcase all manner of moods. The hit single "Fly Me to the Moon" was written in 1954 by Bart Howard and originally titled "In Other Words." Recorded by Frank Sinatra in 1964, it became linked to the Apollo program later in that decade. It was played by the Apollo 10 crew on their Moon-orbiting mission, and was broadcast from Houston Mission Control during the momentous Apollo 11 lunar landing.

Moviegoers made the voyage to the Moon long before NASA's Apollo astronauts. Pioneering French filmmaker Georges Méliès took people on *Le Voyage dans la Lune*—translated as *A Trip to the Moon* for English speakers—as early as 1902. The 1950 film *Destination Moon,* based on a Robert Heinlein novel, won an Oscar for special effects, based in part on the futuristic space art of Chesley Bonestell. In 1968, as Americans watched Apollo missions come closer and closer to reaching the Moon, Stanley Kubrick and Arthur C. Clarke located space travelers—and a mysterious monolith—at a research base within Clavius crater in *2001: A Space Odyssey.*

In short, humanity has enjoyed a millennia-long relationship with the Moon, in myth, folklore, and popular culture, expressed through movies, poetry, artwork, and song. Galileo Galilei, often called the father of observational astronomy, first transformed the inexplicable Moon into an object of scientific study. He reported in 1610 that the dazzling orb had mountains and lowlands. It was not just a featureless lantern in the sky—it was another world! Actually touching down on our closest heavenly neighbor became the goal of generations.

—||—

IT NOW SEEMS A 21ST-CENTURY certainty that human beings will revisit the Moon, although the mission will be far different than those epic small steps taken by the famous first-comers, the Apollo moonwalkers. The Moon of tomorrow is being shaped today by visionary scientists, engineers, designers, and entrepreneurs from around the world. But just as Earth and its sole natural satellite are gravitationally locked in a tense and dynamic relationship, businesses and nations today play tug-of-war relentlessly on matters of politics, economics, and law—interactions clearly influencing the plans and aspirations for exploring and exploiting the Moon and its resources.

President John F. Kennedy's 1961 commitment before the U.S. Congress to land a man on the Moon and return that person safely to Earth was also an unanswered overture to the Soviet Union to join forces and go together. Since then, multiple nations have deployed and landed lunar craft robotically, but no other has placed boots on the lunar surface. Now, more than five decades later, do we see the success of Apollo missions as little more than governmental one-upsmanship, just a dead-ended political sideshow? Was Apollo a future too soon, perhaps a waste of U.S. taxpayer dollars—or was it money well spent and a down payment on the future?

What about the science of it all? Ironically, after billions of dollars have been spent on lobbing robotic mooncraft and propelling humans onto the lunar landscape, theories about the Moon's creation are still debated in scientific circles. The Moon has a tale to tell. It is a record-keeping world rife with patterns of violent change that have transpired in our solar system past, evidence that has been erased on our geologically active Earth but that can be discovered through lunar investigations.

Or perhaps the more energetic discussions about Earth's Moon today will be fed by the competitive spunk between private-sector, profit-making industrialists—discussions that have piqued the interest of outer space lawyers as well. For instance, what laws hamper or promote turning the Moon into a profitable, moneymaking engine? What about the environmental consequences of purging the Moon of its resources?

Orbiters and landers sent to the Moon by many nations have shown an unclaimed wealth of resources. Deposits of water ice might be transformed into oxygen, water supplies, and rocket fuel; it's an alluring mining site, ripe for the picking of rare earth elements. Could these be of international strategic and national security importance? As the Moon harbors very little atmosphere, our celestial cohort has also been viewed as a site for beaming solar energy back to Earth and outward to other destinations in space.

Imagine a future when the Moon tosses off an extra-added glow in the nighttime sky: the visible lights from a lively, sprawling lunar city as humanity acquires long-lasting footing on our nearby world. What will a permanent human presence on the Moon look like? Several architectural firms are already blueprinting lunar domiciles. The European Space Agency has a major initiative to establish a lunar village, an international facility for science, enterprise, and one that acts as training ground for humans to learn how to live on another world, far from Earth. In the meantime, private aerospace groups are hyping tourist flights around the Moon, quick package delivery to the Moon's surface, and lunar modules, including inflatable pop-up living quarters, to form a Moon base.

Consider this futuristic outlook. If preservation advocates are successful, the Apollo 11 Tranquility Base touchdown zone where Neil Armstrong and Buzz Aldrin landed 50 years ago could become the number one tourist hot spot, a mecca for rubbernecking, camera-toting visitors. Encompassed within a large transparent protective dome, onlookers would take in the celebrated landing site, gaze upon still visible footprints stamped in the lunar dust, and witness hardware left behind from that first human sortie to the Moon in July 1969. In this not-so-distant future, Apollo 11's Tranquility Base is just one of many sightseeing attractions, a consequence of nations that cross the void between Earth and our neighboring Moon to image, survey, land upon, and attempt to maintain an enduring foothold on the lunar globe.

After billions of years of slow-going evolution, there is no time out for the Moon. Age aside, things are rapidly about to change.

CHAPTER TWO

WHAT DO WE KNOW ABOUT THE MOON?

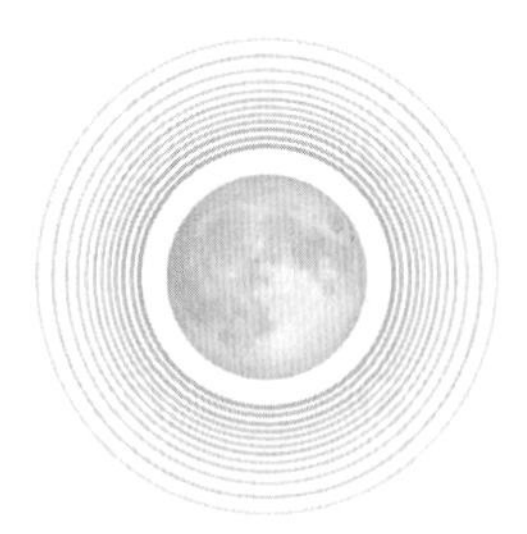

PARADING ITSELF ACROSS THE SKIES, our Moon is a visible spectacle for Earth's 7.6 billion people, an enlightening object available for all to better appreciate the grandeur of the surrounding cosmos. For casual observers here on Earth, there's no need for binoculars or a telescope to find it; the Moon is the most discernible astronomical object in the night sky. It shines brilliantly, but the Moon does not create its own light. Instead, the moonlight we see is reflected sunlight. That light reveals a pockmarked surface, which suggests the Moon has a tale to tell. Even from Earth we can see the Moon's visible surface has two major types of terrain. The "highlands"—also known as the lunar terrae (Latin for "land")—are light-colored and heavily cratered, while the maria (Latin for "seas" and plural of mare) are darker and smoother, formed when lava flows filled depressions on the lunar surface. Surface features visible from Earth also include huge craters, mountain ranges, rills, and lava plains. The Moon's north and south poles host some areas almost always in sunlight and other places always in dark shadow—conditions that play a major role in the future human exploration of the Moon.

Human eyes first spied the far side of the Moon during the Apollo 8 mission in 1968. Astronaut William Anders described the scene outside the spacecraft window: "The backside looks like a sand pile my kids have played in for some time. It's all beat up, no definition, just a lot of bumps and holes." Flybys and lunar orbiters have shown us that the far side has fewer maria than near side.

Prior to early robotic exploration of the Moon, the conventional wisdom was that most lunar craters were volcanic in origin. That view has changed: Studies of the nature of lunar craters themselves and detailed investigations of impact craters on Earth have made it clear that cosmic projectiles have slammed into the Moon. The plainly battered face of the Moon is testimony to the bombardment of asteroids and comets, objects that likely affected early Earth as well. Furthermore, studies of lunar samples returned by the Apollo moonwalkers confirmed the role that impact processes have played in shaping the lunar landscape we see today.

The Moon is "tidally locked" to Earth: The same hemisphere always faces our planet. In short, the Moon has turned its back on us. Despite the misnomer that there's a "dark side" of the Moon, both sides experience two weeks of sunlight followed by two weeks of night. Like the near side, the Moon's far side is also a reserve of geological wonderment. It has one of the largest impact craters in the solar system, the South Pole–Aitken basin. The outer rim of this basin can be seen from Earth as a huge chain of mountains located on the Moon's southern limb, along the edge of the disk we see.

The Moon makes a complete orbit around Earth in 27 Earth days and rotates, or spins, at that same rate, or in that same amount of time. That's why we always see one side. Because Earth is moving as well—rotating on its axis as it orbits the Sun—from our perspective, the Moon appears to orbit us every 29 days. On average, our Moon is 238,855 miles away, although that distance varies because the Moon travels around Earth in an elliptical orbit. Its perigee, or closest approach to Earth, is 225,700 miles. At its farthest distance from Earth, the apogee, the Moon is 252,000 miles away.

Earth's only natural satellite, the Moon is a quarter the size of Earth, making it the fifth largest moon in the solar system. Then there's the gravity of the situation. Because the Moon holds less material than Earth, lunar gravity is less—one-sixth that of our planet's gravity. That means things on the Moon will weigh only 16.6 percent of what they weigh on the Earth: ideal for making giant leaps, as Apollo moonwalkers exhibited when they traversed the lunar surface.

The Moon is not quite an airless orb. It has a very thin atmosphere consisting of helium, neon, and argon, along with smaller amounts of unusual gases not seen in the atmospheres of other rocky worlds, including sodium and potassium, revealing what processes may have created this thin atmosphere. These atmospheric gases could stem from high-energy light and particles from our Sun slamming into the lunar surface and knocking atoms from surface material. Perhaps chemical reactions between that solar wind and lunar surface material caused them, or maybe they are by-products released from the impacts of comets and meteoroids. On the other hand, they might be volatiles that originate from the Moon's interior and break through to the surface, where they sublimate into gas. More study is needed to reveal the nature of the tenuous lunar atmosphere.

The Moon's temperature swings are among the most extreme of any planetary body in the solar system, due in part to this minimal atmosphere. The average temperature on the lunar surface varies based on time of day and location, but at the equator it measures 260.6°F at noon and a super-chilly minus 279.4°F at night. Those temperature swings can make exploration difficult and must certainly be taken into account for future human activities—for example, digging at the surface to collect dust. rock samples, and other material, or to analyze or excavate for building belowground structures.

Below that surface, researchers think the Moon likely has a very small, partly molten core. Scientific lines of evidence imply that the Moon has a solid, iron-rich inner core with a radius of nearly 150 miles and a fluid outer core, primarily liquid iron, with a radius of roughly 205 miles. This conjecture helps shed light on the evolution of the Moon's early magnetic field. Only remnants of that magnetic field remain, isolated in some lunar rock samples. Uncovering and confirming specifics about the lunar core are critical for developing accurate models of the Moon's formation.

A KEY QUESTION for our lunar future is whether there is water on the Moon. For a long time the Moon was commonly regarded as an "anhydrous body"—highly depleted in water and other volatiles. But now, with years of Moon exploration under our belts, we know that the lunar surface environment appears to have exploitable water reservoirs. Although not of any economic value to planet Earth, lunar-derived water may prove essential to establishing a future space-based economy. Of special interest are the craters near the lunar poles with floors in permanent shadow: areas where surface temperatures have been gauged below 40 kelvin—an ultra-chilly minus 388°F. Water ice found within these ever shadowed regions, so-called cold traps, is likely to be stable and usable. Any water ice in such cold traps is likely derived from the impacts of comets or hydrated meteorites with the lunar surface. It is also possible that water or hydroxide molecules are produced by interactions between the wind of particles driven from the Sun and the Moon's regolith—the layer of loose material covering the bedrock. At lower latitudes, these molecules may make their way to the lunar polar regions, becoming trapped there.

The thought of ice in the floors of sunlight-deprived polar lunar craters, called cold traps, was first aired back in the early 1960s. And in the late 1970s, it became vogue to suggest that comets and water-rich asteroids crashing into the Moon could deposit water to the lunar surface. Earth itself acquired its oceans from comets and water-rich asteroids plummeting into our planet.

NASA's Lunar CRater Observation and Sensing Satellite, or LCROSS, added a one-two punch to the investigation of water on the Moon. On October 9, 2009, there were deliberate twin crashes on the Moon, both LCROSS itself and a companion Centaur rocket upper stage that had separated from the LCROSS craft. The target: Cabeus crater, located about 62 miles from the south pole of the Moon, a site almost perpetually in deep shadow due to lack of sunlight. The Centaur craft rushed toward the surface first, with LCROSS following 370 miles behind. These purposeful, powerful impacts of spacecraft on the Moon represented a tandem approach to excavate and spot evidence that the lunar soil within shadowy craters is rich in useful materials. Instruments aboard LCROSS watched as the plume caused by the Centaur craft crash shot up nearly 10 miles above the rim of Cabeus. The spacecraft sent data back to scientists before it also slammed into Cabeus. The headlong hurling of hardware onto the Moon lifted a cloud of lunar matter that may not have seen direct sunlight for billions of years. NASA's Lunar Reconnaissance Orbiter (LRO) observed the crater, debris, and vapor clouds.

After the impacts, grains of mostly pure water ice were detected, lofted into the sunlight in the vacuum above the Moon's surface. The evidence of mostly pure water ice grains in the LCROSS column of tossed-up Moon material gave rise to key questions: How was water delivered to the Moon in the past? What are the chemical processes that cause ice to accumulate in large quantities?

The LCROSS mission also revealed a diversity and abundance of volatiles—chemical materials with low boiling points—in the plume, suggesting an active water cycle within the lunar shadows. In sifting through the data collected by the LCROSS mission strikes on the Moon, researchers determined that as much as 20 percent of the material kicked up consisted of volatiles, including methane, ammonia, hydrogen gas, carbon dioxide, and carbon monoxide, which may have come from a variety of sources, such as comets and asteroids. Also discovered were relatively large amounts of light metals such as sodium, mercury, and possibly even silver. For excited scientists, the LCROSS mission uncovered the Moon's buried treasures: a eureka moment confirming that the permanently shadowed regions of the Moon, cold traps for sure, have been collecting and preserving material over billions of years—materials that in the 21st century can be utilized to sustain human activities.

The Moon is steeped in surprises of late, especially this discovery of water within and on its surface and sequestered at the poles. Moon water is an intense area of research interest and exploration throughout the international community of lunar researchers. It's the most important chemical on Earth, supporting life. And NASA's motto has long been "follow the water"—a saying useful for exploration of the Moon. We now know that these water resources represent fundamental processes active on the Moon. All of these discoveries are thanks to a recent spate of lunar missions; before them, water had been postulated, but unproved.

In 2017, a research team claimed to have found the first direct and definitive evidence for surface-exposed water ice in areas of permanent shadow at the Moon's polar regions. The researchers found specific light signatures reflected off those regions and acquired by NASA's Moon Mineralogy Mapper (M3) instrument, wavelengths of light that correspond to water ice and indicate reserves of water ice absorbing those colors. The M3 instrument flew on board the Indian Space Research Organization's Chandrayaan-1 spacecraft that circuited the Moon between 2008 and 2009. These deposits could fuel future exploration and sustain human outposts. The ice can be melted and distilled to provide potable water, and it can also be broken apart into its constituents, hydrogen and oxygen, to provide breathable air and even make rocket propellant.

Scientists using spacecraft data have also identified widespread water within ancient volcanic deposits, suggesting that the Moon's interior contains substantial amounts of indigenous water as well. The exact origin of water in the lunar interior remains a big question, say Brown University space scientists Ralph Milliken and Shuai Li, who have pieced together the first "road map" of water in the Moon's soil. They found that the signature of water is present across the lunar surface, not solely limited to the polar regions of the Moon. They say that the process of water formation in the lunar soil is active, happening today, and could be a renewable resource, but at what timescale is unknown. Milliken and Li surveyed the upper few millimeters of lunar soil; what the water content is like below that, they aren't sure.

The Moon's inside water story also plays favorably for future lunar explorers; it's an invaluable reserve ready for extraction, perhaps easier to access than water ice in shadowed regions at the lunar poles. After all, no humans have yet set foot at the poles, but they have been to other locales on the surface: All six of the Apollo lunar landing sites were fairly near the equator on the side of the Moon that faces Earth. It remains to be seen whether water ice extraction is feasible at the poles, and whether it makes economic sense to do so. This tantalizing discovery requires more study and on-surface prospecting by robots and humans.

Paul Spudis was a leading advocate for a return to the Moon and senior staff scientist at the Lunar and Planetary Institute (LPI) in Houston. Spudis, who unfortunately passed away in 2018, focused on the geological history and evolution of the Moon and processes of impact and volcanism that have shaped its surface. He proclaimed that there is no question that water exists at the Moon, formed through a long and complex history and several processes. But we still have more to understand. For instance, where does the water come from and what is its chemical makeup? More importantly, what is the true concentration of the water ice and its physical state? Is it just right for easy picking and processing, or hard as concrete?

THE BONE-DRY FACADE of the Moon is now deemed deceiving. What's more, a water-rich lunar viewpoint energizes those eager to return human explorers to the Moon, establish a base camp there, and test technologies that allow humans to rocket off from the Moon to other space targets. Let's say that significant water ice dwells at the Moon's poles. We still need to know the magnitude of those deposits and if they are exploitable. Are those deposits uniform or patchy? Could water ice come in thin layers separated by layers of dust? Proof of the presence of water ice in these regions and the form the ice takes—whether blocks of relatively pure ice or ice crystals mingled in with lunar dust, soil, broken rock, and other related materials—will certainly require further exploration through on-the-spot sampling and analysis, a nontrivial action item. Robotically trekking into the floor of a dark polar crater and digging up a core of whatever is there are both dicey and icy assignments. Once again, the science of discovery could reveal layers of icy comet residue or the leftovers from water-rich asteroids that have struck the Moon over the ages.

David Lawrence, a space scientist at the Johns Hopkins University Applied Physics Laboratory, has reviewed more than five decades of studies concerning polar volatiles on the Moon. He notes that there's a shortfall of measurements by spacecraft circuiting the Moon that are specially equipped to survey volatiles available in permanently shadowed regions—PSRs for short—but PSR research is now blossoming into a thriving subfield of planetary science. Water witching devices toted by lunar orbiting craft could yield information regarding stashes of ice at the Moon's poles. Lunar orbiters can measure the reflections of laser beams shot toward the Moon: If the surface has ice, the laser reflections will be brighter.

To evaluate the nature of PSRs, researchers need on-the-spot measurements, however. A big hitch is operating a landed spacecraft in the super-cold PSR lunar environment. A lunar polar rover can reconnoiter around PSRs, but it's difficult to build machinery able to stay alive during a deep dive into a large PSR, where sunlight doesn't reach. In the years ahead, Lawrence says, there will be a likely upturn in pinning down PSR volatiles on the Moon. Reserves of water ice would be an astronaut's best friend once processed into liquid water, oxygen, and rocket fuel.

Planetary geophysicist Paul Lucey suggests there's a provocative side to cometary leftovers resident within cold traps. Lunar polar ice deposits may be of significant astrobiological interest. The Moon preserves unique historical information about changes in the habitability of the Earth-Moon system, a record obscured on Earth by our planet's active geological processes.

If organics are present in the lunar poles, they represent an opportunity to field-test models of organic synthesis that are proposed for comets and interstellar clouds—processes that could have been important in providing organic material to the early Earth. No other location in the solar system is as convenient as the Moon for this type of investigation, Lucey suggests. The Moon's poles are "luna incognita"—unknown and unexplored territory—that may likely represent a "witness plate"—an accessible, long-duration record of the near-Earth space environment going back to the early history of our solar system. "Ask not what astrobiology can do for the Moon—ask rather what the Moon can do for astrobiology," as Lucey puts it.

Might life have existed on the lunar surface in the distant past? It isn't just a wild idea from one team. In 2018, astrobiologists Dirk Schulze-Makuch and Ian Crawford suggested two early windows of habitability for Earth's Moon, periods in its history that might have been sufficient to support simple lifeforms: First, shortly after the Moon evolved from a debris disk, four billion years ago, and then again during a peak in lunar volcanic activity, around 3.5 billion years ago. During both periods, the Moon was spewing out large quantities of superheated volatile gases, including water vapor, from its interior. This outgassing could have formed pools of liquid water on the lunar surface and an atmosphere dense enough to keep the liquid water there for millions of years.

The earliest evidence for life on Earth comes from fossilized cyanobacteria that are between 3.5 billion and 3.8 billion years old. During this time, the solar system was dominated by frequent and giant meteorite impacts. It is possible that meteorites containing simple organisms like cyanobacteria could have been blasted off the surface of the Earth and landed on the Moon. But whether life ever arose on the Moon, or was transported to it from elsewhere, remains highly tentative and can only be addressed by an aggressive program of future lunar inquiry.

THE MOON IS A MESS. Its whack job face, dinged and gouged from countless, long-ago impacts, is testimony to the cosmic chaos that took place during the early evolution of our solar system. That period of time has been characterized as a massive, madhouse billiard game. Comets, asteroids, moons, and planets were aimlessly slamming into one another for hundreds of millions of years. By studying the shapes, sizes, and ages of lunar craters, researchers gain a window into the early stages of solar system development. It's a telling saga concerning the existence, or nonexistence, of the late heavy bombardment of the Moon around four billion years ago. At the same time, another benefit from crater analysis is spotting hazards and safe havens for future missions, monitoring the Moon for impacts and the risk they pose to future space exploration. Unlike Earth, the Moon has only a feeble atmosphere, so meteoroids have nothing to stop them from making it to the lunar surface.

Case in point: In July 2018, two impacting meteoroids—fragments of asteroids and or comets—hit the Moon with enough energy to produce brilliant flashes of light observable from Earth. These transient lunar phenomena were picked up by the Moon Impacts Detection and Analysis System (MIDAS), a trio of telescopes spread across Spain and equipped with high-sensitivity charge-coupled device (CCD) video cameras that identified the objects hitting the darkened face of the lunar surface. The flashes stood out against the dark lunar ground.

Likewise, NASA's Meteoroid Environment Office in Huntsville, Alabama, monitors the dark portion of the Moon to establish the rates and sizes of large meteoroids (those greater than tens of grams or a few ounces in mass) plowing into the lunar surface. They have cataloged flashes from the Moon, bright light eruptions that stem from the thermal glow of molten rock and hot vapors at an incoming object's impact site. Discoveries of fresh, datable impact craters help to establish the present-day impact cratering rate on the Moon. Hundreds of detectable impacts are taking place every year, and even more go undocumented. Scientists need additional coordinated observations to establish the rate of large meteoroids impacting the Moon.

Another vital impact tool is at work: By comparing photographs from past eras of lunar exploration with the high-resolution data now being relayed by NASA's Lunar Reconnaissance Orbiter (LRO), researchers can see new craters that have formed in the last four decades. A slew of scientists are counting those craters and other surface features that change on the timescale of days. LRO is a potent instrument for researchers to catalog smaller and smaller craters, which dramatically enhances the database about the geology of the Moon.

All these findings contribute to knowledge about how the solar system formed and evolved, says space scientist Clark Chapman of the Southwest Research Institute (SwRI). The Moon is a good place to study those early times, thanks to its minimal recent geological activity, as well as to appraise the first billion years of lunar history. Volcanism and other processes that can now be seen contribute as much as craters to our understanding of the early solar system. Scientists want to know not only *what* events happened, but *when.*

Investigators now have a massive, two-million-crater database of the Moon, which they use to computationally model the crater formation process. One of those researchers, Stuart Robbins, also of SwRI, calculates through the extrapolation process that roughly a half billion to a billion craters created by objects impacting the Moon are greater than 300 feet across, making the Moon a test bed for basic investigations about impact craters. The lunar surface's record of early solar system bombardment gives scientists the ability to look backward in time to see the conditions of early Earth. Distinct from Earth, the Moon did not undergo tectonic plate movements or erosion processes, and has not seen widespread volcanism for billions of years, actions that tend to "erase" telltale geological features like impact craters here on Earth. Without these geological forces, the Moon is left harboring dozens of craters that are roughly a third the length of the United States or larger, Robbins says. By comparison, the largest verified impact crater on Earth—and there are fewer than 200 known, all told, including Meteor Crater in Arizona—is the Vredefort crater in South Africa, which is 190 miles across.

Researchers at the University of Toronto Scarborough have developed a new crater counting technique that uses artificial intelligence. Instead of looking at a lunar image, locating and manually counting the craters and then reckoning how large they are based on the size of the image, observers now can use a software program that automates this entire process, saving significant time and effort. The technique itself relies on a convolutional neural network, a class of machine learning algorithms most commonly applied to analyzing visual imagery. To test out its accuracy, the scientists first trained the neural network on a large data set covering two-thirds of the Moon and then focused the network on the remaining third of the Moon. It worked so well that the new technique was able to make out twice as many craters as traditional manual counting would have done. In fact, about 6,000 previously unidentified craters were found. But the researchers aren't completely satisfied. They're working to improve the algorithm for analyzing visual imagery, to spot more craters.

The procedures of utilizing imagery from lunar orbiters, keeping a sharp eye out for flashes of meteoroid strikes, and applying new computational crater counting tools add up to having a better sense of the Moon's current impact rate. Why is crater counting important? It's not the craters themselves but the impactors that create those craters that researchers want to know more about. Those data are crucial for more effective design of habitats to shield human explorers—and their scientific gear and other crucial equipment—that may reside on the Moon in time to come.

The lunar surface features a stunningly diverse array of volcanic landforms, such as large shield volcanoes, spatter cones, and pyroclastic deposits (rock spewed out from volcanoes)—lunar constructs that look like the small cinder cones and volcanic domes on the Earth. Lunar scientists have long thought that lunar volcanism came to an end about a billion years ago, but that theory is under revision, says Mark Robinson of Arizona State University, the principal investigator for the revered high-resolution camera deployed by NASA's LRO. The interior of the Moon may well be far hotter than once believed, says Robinson, and higher temperatures would have implications for volcanic activity.

Photography taken by the Moon-circling orbiter has led to compelling observational evidence of geologically recent volcanic activity on the lunar surface. Deposits of dried magma terrain—so-called irregular mare patches, or IMPs—scatter the surface in places where volcanic eruptions have left their mark. First spotted in Apollo 15 orbital photography taken in 1971, the best known IMP is called Ina (or Ina-D), an enigmatic volcanic feature known for its irregularly shaped mounds. Robinson explains that these deposits probably formed in the last 100 million years, perhaps even more recently than 50 million years ago. In astronomical timescales, that's just yesterday. But other research published just a few years later says Ina only looks that young; instead, it's about 3.5 billion years old. Perhaps this youthful-looking lunar volcano is masking its true age? More data on this feature are needed.

Ever since the powerful camera on board the LRO has been on duty, starting in 2009, many new IMPs have been discovered across the near side of the Moon. Ina-D is not simply a one-off oddity, Robinson says, but rather a signature of volcanic processes that actually occurred in multiple places. The IMPs not only create a striking landscape; they also speak volumes about the thermal evolution of the Moon, as they are the biggest indicators of recent geological evolution, suggesting that the interior of the Moon may be hotter than previously thought.

The IMPs aren't the only geological structures that tell the story of lunar volcanic activity. Both American and Japanese lunar orbiters have found that the Moon is dotted with caves and expansive lava tubes: sublunarean voids that show themselves as lunar skylights, ranging in size from roughly five yards to nearly a thousand yards in diameter. In recent years, researchers have found more than 200 of these steep-walled pits. Most pits imaged by the orbiters were detected either in large craters with impact melt ponds—areas of lava that formed from the heat of the impact and later solidified—or in the lunar maria, dark areas on the Moon made up of extensive solidified lava flows hundreds of miles across. Scientifically, lunar lava tube studies can help unravel how lava traveled on the early Moon.

Lava tubes are tunnels formed from the lava flow of volcanic eruptions. The edges of a lava flow cool, forming a pipelike crust around the molten river. When the eruption ends and the lava flow stops, the pipe drains and leaves behind a hollow tunnel. Scientists hypothesize that empty lava tubes nearly a mile wide could remain structurally sound on the Moon. If so, considering the much lower lunar gravity, these huge lava tubes could be just right to house an underground city on the Moon. Occupants of subsurface lodging would be shielded from cosmic radiation and the influx of micrometeorites—two major hazards of lunar habitation. To learn more about these tunnel systems, spacecraft designers are discussing a future Moon orbiter outfitted with a radar system to probe beneath the lunar surface, charting the size and shape of lunar caves and their location.

—||—

THE MAGNETIC FIELD OF THE MOON is very weak compared with that of Earth. A planet's magnetic field is generated by what's called a dynamo, formed by the fluid motion of a conducting material, such as liquid iron. On Earth, this motion occurs in the outer core, as liquid iron convects and circulates heat. But the Moon is too small for convection to take place. Scientists now contend that the Moon's solid-rock middle layer, called its mantle, stirs up its liquid iron core. Analysis of magnetized lunar material hauled back to Earth by Apollo missions has shown that the Moon had a strong magnetic field around four billion years ago, and its magnetic field was surprisingly still modest between one billion and 2.5 billion years ago—a billion years later than previously thought. This remnant magnetism suggests that the lunar rock specimens had cooled in the presence of strong fields.

We still have not pieced together the full story of lunar magnetism. Ongoing research, including acquiring new lunar samples from different locations on the Moon by way of robotic craft, is in the offing. Added insight about how the Moon's dynamo originated and developed helps to understand dynamos of other worlds, like planets and other moons. Because a magnetic field shields the atmosphere of a planet or moon, and the atmosphere protects the surface, together these two heighten the potential habitability of a planet or moon, perhaps even on worlds far beyond our solar system.

These remnants of a global field aren't the only hints of magnetic activity on the Moon. The lunar surface is decorated with features called "lunar swirls"—odd tattoo-like blemishes that can be tens of miles across, appearing in groups or isolated. Lunar swirls resemble bright, snaky clouds painted on the Moon's dark surface. One of the most studied swirls, about 40 miles long, is Reiner Gamma in western Oceanus Procellarum.

Scientists often spot lunar swirls where ancient bits of magnetic field are also found embedded. They note that the bright areas in the swirls look to be less weathered than their surroundings. Ever since identifying these swirls in the 1960s, researchers have debated what they are. Most lunar swirls share their locations with powerful, localized magnetic fields, so they appear to be related to magnetism, but the cause of those magnetic fields and the swirls themselves is still in question.

Observations of these strange features from Earth and via Moon orbiters have led to theories about how lunar swirls form. The bright and dark patterns may result when those magnetic fields deflect particles from the solar wind and cause some parts of the lunar surface to weather more slowly. Or, because particles (electrons and ions) in the solar wind are electrically charged, perhaps those particles respond to magnetic fields and land in particular patterns on the Moon, suggesting that the magnetic field shields the surface from weathering. Perhaps these swirls could have formed from plumes of material ejected by comet impacts. Another theory purports that when micrometeorite impacts loft fine lunar dust particles, an existing magnetic field at those locations sorts them according to their susceptibility to magnetism, forming light and dark patterns with different compositions.

A joint study on lunar swirls was issued in July 2018, led by researchers at Rutgers University and the University of California, Berkeley, and suggesting that the swirls may have resulted from ancient magnetic lava just below the lunar surface that is tied to long, narrow lava tubes or lava dikes, or vertical sheets of magma in the lunar crust. The high iron content in lava tubes or dikes would have become strongly magnetic as they cooled, the researchers concluded—the most likely scenario for how magnetism underlies these lunar swirls. A surface mission to one of these features could probe many of the questions swirling around swirls. Magnetic anomalies are natural laboratories as they offer a venue to examine sets of key issues that bring together the science of the sky above (solar ultraviolet flux, solar wind properties, and energetic particles) with the geosciences underfoot, says David Blewett of the Johns Hopkins University's Applied Physics Lab.

For instance, a robotic rover driving across a swirl, says Blewett, could measure how much solar wind is striking the lunar surface, investigating what causes the soil and swirls to darken with time. Ideally, a rover could peg the origin of magnetic anomalies on the Moon. Resolving the true nature of the swirls, their occurrence on the lunar surface, and why they are related with magnetic anomalies will help shape our understanding of the history of the Earth-Moon system, the interaction of planetary surfaces with the solar wind, and how we can fruitfully explore planetary surfaces.

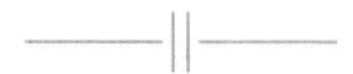

THE REAL PUZZLER IS: Where did our Moon come from? It's an innocent, childlike question, but today those are six words that engender debate, scads of peer-reviewed geological papers, and downright disorder in scientific circles. Toss in the fact that even with $25 billion spent on Project Apollo so that moonwalking crews could scrutinize the scene up close, set up scientific gear, and collect lunar samples, we still have only clues but no definitive answer as to the precise origin of our cosmic companion.

Perhaps our Moon formed elsewhere, prior to being caught in Earth's gravitational pull? Maybe the Moon co-formed with Earth, condensing from gas and dust 4.5 billion years ago? Or was the Moon a spin-off, the product of a large object that smashed into Earth as our planet was in its formative stages? Bandied about over time are the pros/cons for these and other concepts. Whatever truth is finally revealed, that truth will be—to abuse an old cliché—"Something old, something new, something borrowed, something *blew!*"

The source of the Moon within the Earth-Moon system is one of the most basic questions in planetary and lunar science. The conventional hypotheses fall into four categories: the fission theory, the condensation theory, the capture theory, and the giant impact theory.

The "fission theory" holds that the Moon was once part of Earth and somehow separated from our planet early in the history of the solar system. This hypothesis suggests that the present Pacific Ocean Basin is the location from which the Moon came from. The idea was thought feasible because the Moon's composition resembles that of the Earth's mantle, and a rapidly spinning Earth could have cast off the Moon from its outer layers. But one problem is that the present-day Earth-Moon system should contain "fossil evidence" of this rapid spin, and it does not. Another problem is the comparative ages of these two features: The Pacific Ocean is less than 200 million years old, whereas the Moon formed 4.5 billion years ago.

According to the "capture theory," the Moon formed somewhere else in the solar system and was later pulled in by Earth's gravitational field. Capturing the Moon into its present orbit seems dubious, however. The Moon and Earth would have had to pass each other at low speeds, while the Moon's orbit lost energy as it approached Earth. Those are two very specific details, and scientists are reluctant to believe in such "fine-tuning." Furthermore, Earth would not have had enough energy to gravitationally imprison the Moon, and so it would have been released at some later point in time—which obviously hasn't happened.

The condensation theory, also called the co-accretion theory, posits that the Moon and Earth formed at the same time, with the Moon in orbit around Earth. Research suggests the Sun, the planets, and all other objects in the solar system formed from a cloud of gas and dust, a nebula that coalesced billions of years ago. If the Moon was created in the vicinity of Earth, though, it should have nearly the same composition. They would share similar properties, such as material densities, and the makeup and size of their inner cores. And because Earth has more mass than the Moon, if the two had condensed simultaneously, Earth's gravitational force would have pulled in the Moon to become part of the planet and not a natural satellite.

The giant impact theory is the favored formation theory among many scientists. According to the early version of this idea, a Mars-size object, dubbed "Theia," collided with our

Earth 4.5 billion years ago. (The name comes from Greek mythology: Theia was the mother of the moon goddess Selene.) This slapdash celestial kiss between worlds flung vaporized rock and debris from both bodies into space. Some of it went into orbit around Earth and then congealed. From the resulting debris of this car crash of astronomical proportions, our Moon became Earth's only natural satellite. More recent work suggests we may need to revise our sense of the size of the proto-Earth and its giant impacting object.

The giant impact theory has emerged as the leading explanation for how the Moon came to be, but most of the evidence results from computer modeling and theory, not observation. Physical evidence to decipher formation models is not easy to come by. Modeling efforts largely use the giant impact hypothesis as the basis of their studies, bolstered by research that suggests large impacts were quite common events in the late stages of terrestrial planet formation. Moon samples brought back from Apollo missions do bolster this theory, sporting characteristics that match Earth rocks. Matching rock is the geological equal of matching fingerprints.

But science never sleeps. Theories and hypotheses must always be painstakingly mulled over until the right one advances and is universally accepted. Here's a captivating question for the giant impact Moon camp. If the Moon formed early in solar system history from debris left behind after an object careened into Earth, the Moon's interior should be dry, as it appears doubtful that hydrogen could have endured the heat of that impact. But scientists have detected signs of water, and water requires hydrogen. How do those discoveries fit into the Theia model?

Indeed, the giant impact hypothesis is far from certain. Unearthing the true origin of the Moon is an iterative, methodical process requiring repeated rounds of analysis, questioning every hypothesis. Some, like lunar geologist Harrison Schmitt, suggest that the Moon accreted independently, roughly in the same orbit as the Earth, and then was captured. As evidence, they point to a striking feature in the compositions of the Earth and Moon: They have identical abundances of the isotopes of oxygen and other elements. Isotopes are atoms that have the same number of protons and electrons but different numbers of neutrons and therefore have different physical properties. If Earth and the Moon have identical abundances of isotopes, they could not have come from significantly different regions of the solar system. And it's difficult to explain how an incoming body slamming into a young Earth, throwing extremely hot material off the young planet that then condensed, could ensure that both objects have the same isotopes, as well as leave major reservoirs of volatile elements deep within the Moon.

Work funded by the NASA Lunar Science Institute in late 2012 suggested another formation theory, hypothesizing that our early Earth and Moon were both created together in a giant collision of two planetary bodies that were each five times the size of Mars. Although this is still an impact theory, the involved bodies were very different from the young Earth and the Mars-size Theia. This celestial scenario envisions a primary collision between the two bodies, followed by a recollision of the objects, forming an early Earth encircled by a disk of material. Those bits and pieces later pooled and cooled to make our Moon, circumstances leading to an Earth and a Moon with similar chemical compositions.

The new theory rocks the boat, challenging the commonly believed giant impact paradigm including a "smallish" impactor the size of Mars. Robin Canup of the Southwest Research Institute in Boulder, Colorado, developed this newest theory. In her computational simulations, which are based on special equations of motion, each of the clashing planets is divided up into about a million particles, and scientific modeling tracks characteristics of each particle, including position, velocity, temperature, and gravitational interactions during the collision.

In 2017, another team of scientists challenged the idea that the Moon was born by way of a single giant collision, citing inconsistencies with this scenario. Researchers at Israel's Weizmann Institute of Science argue that a multiple-impact origin for the Moon makes more sense. After hundreds of computer simulations, the research team reported that a "more plausible chain of events" might involve a number of run-ins with smaller objects, yielding smaller moonlets that crossed orbits, collided, and merged—a long series of moon-moon rear-enders that gradually built up a bigger moon. Those collisions involved objects ranging from small planets one-tenth the mass of Earth to space rocks the size of the Moon, one-hundredth the mass of Earth. The cosmic smashups would have sent clouds of rubble, melt, and vapor into orbit around the early Earth. Some collisions would throw off a bit of material that would eventually combine with another bit of material; some collisions wouldn't throw any rocks off. Hundreds of simulations by Raluca Rufo and colleagues suggest that the debris, over time, would cool and agglomerate into small moonlets that could have combined into a larger moon—our Moon today.

These new models show that our understanding of the solar system is continually being refined, and exemplify the role of modeling planetary formation in our search for understanding just how our Moon came to be. Scenarios for Moon formation will continue to be assessed and reassessed in combination with ever improving models of terrestrial planet formation. The best way to test these models is to collect new samples from the Moon to better assess the chemical composition of the Moon and determine when any giant collision occurred. Although the Apollo missions were extraordinarily successful, they explored only a small fraction of the lunar surface and were therefore unable to collect the best samples for answering the question about the Moon's origin.

The Moon's rocky face, along with the rest of the terrestrial planets, hint at a time of unrest in the solar system. In fact, for the past 45 years, scientists have generally agreed that some 3.9 billion years ago, a sudden 150-million-year spike in bombardment formed most of the major basins on the Moon. The theory is that such basins formed because the intense heat caused by such impacts actually melted the rock that was struck, and no impact melts seemed evident before 3.9 billion years ago. The same thing would have happened on Earth, but our planet's changing geology has covered those battle scars. Scientists call this period the late heavy bombardment, also known as the lunar cataclysm.

But some scientists are critical, and this model is beginning to collapse after being the lead theory for decades. William Hartmann, senior scientist emeritus at the Planetary Science Institute in Tucson, Arizona, suggests that late heavy bombardment observations are best explained not by a lack of Moon impacts before four billion years ago. Rather, a very high impact rate before four billion years ago rapidly chewed up evidence on the lunar surface. Large impacts created shock melt and vaporized material that later lined the cavity of craters. After formation, this shock melt pooled into a sheet of molten material that makes up the floors of large craters. Then, about 3.9 billion years ago, he adds, the impact rate dropped to a level where many impact melts survived. That's why data today suggest a spike at 3.9 billion years ago.

Hartmann's historical review project, ongoing at the International Space Science Institute in Bern, Switzerland, is not so much about adding new science as it is about determining how scientific ideas are started and why people believe them. In the case of the late heavy bombardment, he asserts each group thought the others had proven the concept, so that the edifice was constructed from data and models that were nonconclusive.

Our understanding of the solar system is continually being refined. New models exemplify the importance of modeling as a technique to discover how our Moon came to be. We cannot go back in time to watch what happened. Instead, scientists create computer models that line up with the physical evidence—and there's only so much physical evidence. The scenarios for Moon formation will need to be assessed and reassessed by ever improving models of terrestrial planet formation. Researchers will also make good use of samples from the Moon's surface and belowground.

Future exploration of the Moon builds upon decades of early pioneering space missions that were foundational in divulging the past history of our companion world. A variety of signature spacecraft were instrumental in progressively setting the stage for how investigations of the Moon will go forward in decades to come. There have been stunning advances in lunar science over the past decade, but many compelling scientific questions remain. Earth's Moon has not disappointed in showcasing its complexity and perplexity. It is a spectacular, close-up world full of wonder and opportunity. It beckons us—humans—to explore its inner and outer workings

CHAPTER THREE

PAVING THE WAY

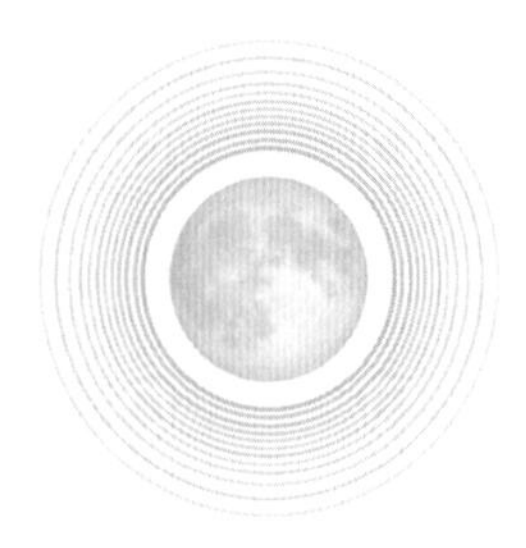

IN OCTOBER 1957, Earth acquired a shiny artificial moon in the form of a 184-pound polished sphere called Sputnik 1. That landmark occasion broadcast more than just beeps. It sent technological, scientific, and political shock waves around the globe. Rocketed into orbit by the Soviet Union, Sputnik 1 was a sobering escort of Earth. That diminutive sphere acted as a warning sign to the United States; it accelerated Cold War tensions, representing a step up in military weaponry development. Since the late 1940s, the United States and the Soviet Union (now disassembled into several countries, Russia the largest) had been locked in an escalating war of ideas—with few weapons fired. Relations between the United States and the Soviet Union were unmistakably edgy, a superpower rivalry fueled by differences in political ideology and economic pursuits, with both nations striving to influence the world by showcasing their technological and military prowess.

But because of those tensions, the Earth-circling Sputnik vividly verified that Russia had the booster power to hurl a warhead on a ballistic trajectory to any point on the planet. As American historian Daniel Boorstin reflected: "Never before had so small and so harmless an object created such consternation." With Sputnik serving as a slap in the face, the following years saw the shoot-down of an American U-2 spy plane in Soviet airspace and the Cuban missile crisis, which brought the two countries to the brink of war. Sputnik 1 had another effect: It sparked the "space race." A fierce, competitive contest had begun. Sputnik galvanized unexpected governmental change in the United States, and led to the 1958 creation of NASA. The ultimate goal—for both countries—was not purely to send a metal sphere around Earth, but instead to launch humans into space and to attain the wherewithal to head for the Moon. To achieve that objective, robotic craft first had to make the voyage. The starting gun had been fired and the finish line was the Moon.

Decades after Sputnik, space achievements remain as credible signals of national technical, economic, and organizational aptitude. Even now the Moon remains the end point for robotic spacecraft from many nations, sent there by America, Russia, India, China, Japan, and Europe. From the Soviet Union's earliest flybys of the Moon to the kamikaze-like crash landings of America's Ranger probes, the lunar landscape came into focus in the age of Sputnik. These surrogate sentinels of exploration transformed our knowledge about the Moon—and continue to do so.

—||—

JUST 15 MONTHS ELAPSED between Sputnik 1's 1957 launch and quest for the first craft to reach lunar territory. In January 1959, the Soviet Union's Luna 1 began a series of flights of Soviet interplanetary stations sent toward the Moon. Luna 1 was a hermetically sealed sphere, the first spacecraft to achieve a velocity fast enough to escape Earth's gravitational pull. Even though Luna 1 missed the Moon and went into orbit around the Sun, it did accomplish a science first: It charted the wind of particles coming from the Sun.

But the Soviet Union was intent to crash a craft on the Moon—and achieve a political punch by delivering two metallic pennants with the U.S.S.R.'s coat of arms onto the lunar surface. On September 13, 1959, the Soviet Union achieved that goal with Luna 2. Later that same year, Luna 3 circled the Moon, providing the first views ever of the side never visible from Earth. It was thanks to the "early, poor-quality images" from Luna 3, as lunar scientist Paul Spudis put it, that we learned that "the far side has surprisingly little of the dark, smooth mare plains that cover about a third of the near side" and is instead densely peppered with impact craters of every size and age.

As Soviet lunar firsts were stacking up, the United States was embarrassingly haunted by failures, showing itself unable even to send simple probes to fly by the Moon. In March 1959, NASA launched Pioneer 4, the first American spacecraft to travel fast enough to escape Earth's gravity. But it missed the Moon by more than 37,200 miles.

For the moment, the Soviet Union continued to excel at space exploration. In April 1961, they put the first man in space: cosmonaut Yuri Gagarin, who spent 108 minutes orbiting Earth. On May 5, 1961, U.S. astronaut Alan Shepard flew a suborbital flight of 15 minutes, 28 seconds. That test—a quick, albeit highly televised event—set America's human spaceflight program in motion.

Just 20 days after Shepard's short, suborbital, but trailblazing trek, President John F. Kennedy stood before Congress and set America's trajectory for the Moon, an initiative that would later be named Project Apollo. The young president challenged NASA and the nation to "land a man on the Moon and return him safely to Earth" before the decade of the 1960s ended.

President Kennedy's rallying cry for the United States to reach for the Moon was amplified in his famous speech, delivered on September 12, 1962, to 35,000 people at Rice University. "But why, some say, the Moon? Why choose this as our goal? And they may well ask why climb the highest mountain? Why, 35 years ago, fly the Atlantic?" Kennedy outlined in matter-of-fact style the difficult issues that had to be tackled. "We choose to go to the Moon in this decade and do the other things, not because they are easy, but because they are hard, because that goal will serve to organize and measure the best of our energies and skills, because that challenge is one that we are willing to accept, one we are unwilling to postpone, and one which we intend to win."

The goal was daunting. The young president stressed the technical test ahead, challenges that only determination and daring could overcome. "But if I were to say, my fellow citizens," Kennedy stated, "that we shall send to the Moon, 240,000 miles away from the control station in Houston, a giant rocket more than 300 feet tall, the length of this football field, made of new metal alloys, some of which have not yet been invented, capable of standing heat and stresses several times more than have ever been experienced, fitted together with a precision better than the finest watch, carrying all the equipment needed for propulsion, guidance, control, communications, food and survival, on an untried mission, to an unknown celestial body, and then return it safely to earth, reentering the atmosphere at speeds of over 25,000 miles per hour, causing heat about half that of the temperature of the Sun—almost as hot as it is here today—and do all this, and do it right, and do it first before this decade is out, then we must be bold."

Kennedy set the bar high, but in this moment the Apollo program, which would ultimately take the first human beings to the Moon, began to take shape. The United States needed to achieve many steps to prepare for a crewed mission to the Moon. These precursor steps involved both robotic spacecraft and the first crewed crafts. Robotic precursors would gather valuable information, constituting the first scientific exploration of another planetary body. To make sure that human crews could safely land on and depart from the lunar surface, it was critical to understand the lunar environment, its surface and processes—and robotic craft performed that priceless duty.

AMERICA HAD ITS FIRST SHOT at taking the lunar lead in the space race, with the Ranger hard lander program. The camera-carrying vehicles in the Ranger series were designed to image the lunar surface at increasing levels of detail on the way to intentional self-destruction as each hurtled into the Moon's surface. The Ranger photographs provided valuable information for future landing site selection, used in the subsequent Surveyor soft-lander program, and documented the types of lunar terrain that Apollo landing missions could visit.

The Ranger program began in 1961, but encountered a chain of excruciating calamities. Booster failures, spacecraft misfortunes, and moon misses dogged the first six missions. All that anguish spurred Congress to initiate an investigation into "problems of management" at NASA Headquarters and the Jet Propulsion Laboratory (JPL).

"At its outset, the project's priorities were driven by the post-Sputnik urgency of competition with the Soviet Union," recounts James Burke, then a beleaguered JPL project manager of Ranger. Ranger 6 had a perfect flight to the Moon, but unfortunately its video transmitters suffered a technical glitch early in the mission. Due to a failure of the camera system, no images were returned. "Ranger 6 was the one that drove everyone crazy," Burke remembers.

The program was reorganized, a new project manager stepped in, and what followed was near success. Six months later, Ranger 7 succeeded in sending back thousands of detailed television pictures of Mare Nubium—the Sea of Clouds. In February 1965, Ranger 8 swept an oblique course over the south of Oceanus Procellarum and Mare Nubium, finally crashing in Mare Tranquillitatis, about 43 miles from where Apollo 11 would land four and a half years later. The Ranger program culminated in March 24, 1965, with Ranger 9 careening into the stunning lunar crater Alphonsus.

All in all, the Ranger program was a success. Ranger's images provided better resolution than was available from Earth-based views of the Moon by a factor of 1,000 and were far sharper than those Soviet Moon probes obtained. From the set of Ranger missions, we discovered the size of craters that pepper the lunar surface, ranging down to the very limits of resolution. Those detailed images also revealed a sobering reality: Finding smooth landing sites for spacecraft carrying humans was not going to be easy.

—||—

DURING THE FIRST YEARS OF NASA, as the civilian space agency was developing the Ranger program, branches of the U.S. military had a few lunar plans of their own. The space race was at its core a duel between the United States and the Soviet Union: Which country was technologically superior, who had the capability to launch targeted attacks, but more importantly, who had the ability to control the new frontier of space? If NASA couldn't get it done, America would find another way; —military proposals were on the drawing board at the same time as the NASA plans that actually came to fruition.

Those proposals are outlined in documents from the U.S. Army and Air Force recently made public. In July 2014, on the 45th anniversary of the first human landing on the Moon, the National Security Archive at the George Washington University released declassified documents that reveal the covert side of lunar programs under way in the United States and the Soviet Union. Although the U.S. civilian program to reach the Moon and some details of the Soviet program were public, other aspects of the race were kept secret, as Jeffrey Richelson explained in his release of the documents. A senior fellow with the National Security Archive, Richelson was a leading academic expert on the process of intelligence gathering and national security.

These previously classified records show how attuned the U.S. military was to space-based activities, including the possibility of nuclear tests in space, the use of the Moon to reflect signals for military or intelligence purposes, and U.S. intelligence analyses and estimates of Soviet missions and their intentions to land a man on the lunar surface. Even more telling, they show that prior to NASA, a civilian agency, being given the task of landing a man on the Moon, both the Army and the Air Force lobbied to establish outposts there. The Army's plan argued for a military Moon base that would be used to develop techniques for surveillance of both Earth and space, communications relays, and other jobs on the lunar surface, taking into account political, management, policy, and legal implications. "To be second to the Soviet Union in establishing an outpost on the Moon would be disastrous to our nation's prestige and in turn to our democratic philosophy," stated the Army study.

The Air Force put forward two separate plans, one to start in November 1964 and to complete an operational lunar base by June 1969—a month ahead of the history-making Apollo 11 Moon landing that actually happened. The second Air Force study spelled out a first manned landing and return in late 1967, part of a master plan leading to a permanently manned lunar outpost in early 1968. "This one achievement, if accomplished before the U.S.S.R., will serve to demonstrate conclusively that this nation possesses the capability to win future competition in technology," read the Air Force document. "No space achievement short of this goal will have equal technological significance, historical impact, or excite the entire world."

NASA FOLLOWED THE RANGER EFFORT with the highly successful Surveyor program shortly thereafter. Seven robotic spacecraft, launched between May 1966 and January 1968, were built to assess the Moon's surface conditions and validate the technology for the soft touchdowns that would be essential for human landings. They also added to overall scientific knowledge about the Moon.

Each Surveyor was outfitted with a television camera. By relaying close-up images of the lunar surface and performing the first soft landings on the Moon, the Surveyor program taught mission planners how to safely set down on the barely explored terrain. Overall, five robot Surveyor vehicles landed safely on the Moon; two others were lost in their attempts to touch down. Four of these examined widely separated mare sites in the Moon's equatorial belt, sites being considered for manned lunar landings. The fifth investigated a region within the southern highlands close to Tycho crater, a site chosen primarily for its scientific interest.

Two of the Surveyor landers carried a surface sampler scoop to assess the Moon's surface. The instrument gouged out trenches in the lunar topside, assessed its bearing strength, and otherwise manipulated the lunar material in view of the television system, conveying images back to Earth for analysis. Surveyor 6 acquired 30 hours of data on the chemical composition of the lunar material. Four spacecraft survived the extreme cold of the lunar night and operated for more than one day/night cycle. In total, the five spacecraft operated for a combined elapsed time of about 17 months, transmitted 87,000 pictures, performed six separate chemical analyses of surface and near-surface samples, dug into and otherwise manipulated and tested lunar material, measured its mechanical properties, and obtained a wide variety of other data that greatly increased our knowledge of the Moon.

Surveyor analysis of the lunar surface found that the dark maria of the Moon are similar in composition to terrestrial basalt, a dark iron-rich lava. The lunar highlands near Tycho, on the other hand, are lighter in color and rich in aluminum—a chemical element that came as a surprise to researchers, because the mare sites investigated by previous Surveyors had not shown a high abundance of aluminum. Apollo moonwalkers later returned lunar samples that confirmed these distinctions.

At the same time as the Surveyors were probing the lunar surface, NASA pushed forward on its robotic reconnaissance of the Moon with the Lunar Orbiter program. From August 1966 into August 1967, five Moon-circling craft relayed photography of most of the Moon, both near and far side. Those high-resolution pictures, which could spot boulders as small as a couple of meters across, helped select and certify safe areas for landing crews. Lunar Orbiter imagery produced first-time "pilot's-eye" views of what Moon landing crews would face as they approached the surface. The first three missions were dedicated to imaging 20 potential lunar landing sites; the fourth and fifth missions were devoted to broader scientific objectives. Lunar Orbiter 4 photographed the entire near side and 95 percent of the far side, and Lunar Orbiter 5 completed the far side coverage and acquired medium- and high-resolution images of 36 preselected landing areas.

All in all, these preparatory U.S. robotic missions, both Surveyor on the surface and Lunar Orbiter overhead, found the

Moon to be cratered and pitted at all scales, carpeted with a powdery dust, and hosting a surface strong enough to take the weight of robotic vehicles and eventually humans. In short, NASA's programs greatly illuminated Apollo's way.

AT THE SAME TIME, the Soviet Union had its share of successes in landing robotic craft on the Moon. In fact, the U.S.S.R.'s Luna 9 was the first spacecraft to pull off a lunar soft landing. In early 1966, using a retrorocket and air bags, the craft landed—albeit bouncing and rolling to a full stop—on the Moon's near side Oceanus Procellarum, the largest of the lunar mare. After the probe came to rest, four petals on the top shell of the spacecraft opened outward and stabilized the spacecraft on the foreboding terrain.

For more than eight hours, Luna 9 captured a series of TV pictures and radioed them to Earth. When the photos were pieced together, panoramic style, the result was an impressive view of the nearby lunar surface and the horizon far away from the lander. Those images also showed the lunar surface as powdery and strewn with rocks, yet strong enough to take the load of a landed spacecraft. Luna 9 also returned radiation data before its battery power was depleted, ending its three-day historic mission.

Even in early 1967, the U.S.S.R. maintained its push to beat the United States in the race to the Moon, planning a piloted circumlunar flight before the 50th anniversary of the Communist state that November. A series of robotic Zond spacecraft were sent on circumlunar flights into the early 1970s, in the hopes that a piloted Soviet Moon program would soon follow. The most successful of these, Zond 5, deployed in September 1968, carried the first life-forms from Earth around the Moon and back again. Two tortoises, worms, flies, seeds, plants, and bacteria all made the round-trip journey and successfully splashed back down on Earth in the Indian Ocean.

In November 1970, the U.S.S.R. also landed the first successful robotic lunar rover, a solar-powered machine that wheeled across the Moon during the day, then parked and relied on thermal energy from a polonium-210 radioisotope heater to survive during the bitter cold nights. Operators back in Russia drove the *Lunokhod 1* rover for 10 months. It traveled more than six miles within the Moon's Mare Imbrium, loaded with scientific instruments that transmitted back to Earth masses of information important for future Moon exploration. The *Lunokhod* relayed data concerning the composition of the Moon's face, took close-up views of the local topography, made video panoramas, and provided engineering measurements about how well a wheeled vehicle traversed the lunar regolith.

In January 1973, an upgraded *Lunokhod 2* landed safely and started scooting across the Moon. This robot had higher-resolution cameras and an improved scientific payload. As with its predecessor, engineers on Earth drove the rover during the day and parked it at night. *Lunokhod 2* scouted the Moon scenery for about four months, chalking up more than 24 miles of travel—a distance verified by a NASA orbiter in 2013.

Driving lunar rovers from a world far away was not easy. Radio signals travel at the speed of light, but the distance between the Earth and Moon means the signals take 2.6 seconds per round-trip—a characteristic called two-way "latency." Network delays slowed the signals even more, amounting roughly to a 10-second latency. These circumstances made operating the *Lunokhod*—driving it around and avoiding obstacles in its path—difficult and mentally challenging. Imagine what it would be like driving your car through a parking lot with a 10-second delay between your eyes and your hands and feet!

To what extent were the *Lunokhod*'s precursors of what may happen on the Moon in years to come? Three of the Apollo human landing missions incorporated rovers on the Moon, although those craft had humans sitting in the driving seat. These vehicles were popularly known as "Moon buggies."

Unwilling to lose the race to the Moon competition with the United States, the Soviets did indeed execute an impressive trio of sample return missions in the 1970s. Their robotic Luna missions successfully autopiloted soft landings on the Moon and then collected samples, rocketed them back, and parachuted them safely back to Earth—a notable engineering feat in its own right. The U.S.S.R.'s Luna 16 returned with a 101-gram sample from Mare Fecunditatis in September 1970; Luna 20 returned 55 grams of soil from the Apollonius highlands region in February 1972; and in 1976, Luna 24 retrieved 170.1 grams of lunar samples from another mare. This was the last Soviet mission to the Moon.

—||—

ROBOTIC SPACECRAFT TOOK PICTURES and measurements, collected samples, and provided vast amounts of information about the lunar surface—but learning about the Moon was just part of the plan. NASA's mission was also to launch humans safely into space and support them during the multiday trip to the Moon and back. The United States developed several programs, each building off the previous one, to prepare technology for human spaceflight. Apollo itself was progeny to a string of stepping-stone space missions.

The Mercury program began officially in November 1958, its goal to put American astronauts into orbit around Earth. Those were the first U.S. astronauts, and they were chosen to fit specific characteristics: flight experience, ability to navigate stressful circumstances, and willingness to take orders. The people who fit these criteria the best, according to mission planners, were test pilots within the armed services.

Between May 1961 and May 1963, six astronauts flew aboard Mercury capsules, the opening volley in America's human space program. As NASA's inaugural spaceflight program, Project Mercury had central objectives: to orbit a manned spacecraft around Earth to investigate the human ability to function in space, and to develop the ability to recover both vehicle and passenger safely back on Earth. Beginning with Alan Shepard on May 5, 1961, six Mercury missions launched successfully.

NASA's fast-paced Gemini program followed the Mercury missions. Ten two-person crews flew Gemini spacecraft during just two years, 1965 and 1966, the link between the earlier Mercury and the soon-to-come Apollo programs. Whereas Mercury missions were valuable in showing that the United States could put astronauts in space, Gemini was a skill-shaping undertaking. Gemini missions included rendezvous and docking of spacecraft, extended the time crews lived and worked in microgravity, and in some cases involved space walks. The Apollo program would demand all of these and more.

The two-seater Gemini missions launched atop Titan boosters. Their space sojourns lasted for periods ranging from five hours to 14 days. The general objectives of the program included: Achieve long duration flights; evaluate the ability to maneuver a spacecraft through rendezvous and docking of two vehicles in Earth orbit; perform onboard orbital navigation; conduct experiments in space; carry out extravehicular operations (space walks); and actively control reentry for a precise landing. That impressive list of accomplishments rocketed the United States into the lead during the Cold War "space race," but the daunting and daring challenge of sending humans to the Moon loomed large, driven by meeting Kennedy's declaration that the United States would land astronauts on the Moon by the end of the 1960s.

To meet that objective, the space agency had to develop not only the modules that astronauts would fly in but also the propulsion technology to launch and carry those heavy pieces of hardware into a lunar trajectory. To land a man on the Moon by the close of 1969, they had to build and test powerful rockets. The Saturn rocket boosters provided the lift and power, and the potent Saturn V proved able to send crews out of Earth orbit and moonward. Tests proceeded at a fast pace and included unmanned Saturn I missions, pad abort tests of a launch escape tower, as well as uncrewed suborbital and Earth-orbiting missions, all focused on contributing to successful Apollo missions carrying man to the Moon by 1969.

—||—

UNFORTUNATELY, THE APOLLO crewed flights began with tragedy. On January 27, 1967, during an on-the-pad test with the intended first crew of astronauts, a fire swept through the Apollo command module. It claimed the lives of astronauts Virgil Grissom, Edward White, and Roger Chaffee. That devastating day portrayed the reality that pioneering the space frontier was indeed a dangerous enterprise. The Apollo program was put on hold during an intensive and introspective investigation of the accident. It was found that the most likely cause of the horrifying mishap was a spark from a short circuit in a bundle of wires within the spacecraft. The oxygen-rich environment in the cabin contained so much flammable material, the fire spread quickly, taking the lives of the trio of astronauts within minutes.

The AS-204 mission was redesignated Apollo 1 in honor of the fallen astronauts. The investigation resulted in major design and engineering modifications in the Apollo program. After the accident, a set of unmanned flights preceded Apollo 7, October 1968, the first piloted flight and shakeout of the Apollo spacecraft in Earth orbit.

Three astronauts could travel aboard the Apollo spacecraft, which itself was made of three sections, the command module (CM), the service module (SM), and the lunar module (LM). The command module held the crew's quarters and flight control section. The service module consisted of the propulsion and spacecraft support systems (the command and service modules remained connected, and were collectively called CSM). The lunar module would transport two of the crew to the lunar surface, support them on the Moon, rocket them back to the CSM, which was to remain in lunar orbit. With all three astronauts aboard, the CSM would cast off the LM and the threesome would then journey back to home planet Earth. A heat shield on the Apollo spacecraft protected the crew during the high-speed return through Earth's dense atmosphere. A trio of large parachutes was designed to unfurl to cushion the crew for their watery landing and recovery.

The primary objectives for the October 1968 Apollo 7 engineering test flight were straightforward: Trial run the command and service modules, evaluate crew performance, and demonstrate the rendezvous capabilities of the new spacecraft systems. Apollo 7 spent more time in space than all the Soviet piloted spaceflights combined up to that time. The crew orbited the Earth for nearly 11 days and cleared the way for an audacious lunar orbit mission to follow. Just traveling to the Moon would take three days, and a full-length mission where astronauts landed on the lunar surface and then returned to Earth would span even longer. Apollo 7 showed that the program could support humans in space for a prolonged length of time, compatible with back-and-forth travel to the Moon.

The next mission, Apollo 8, lifted off on December 21, 1968, and it was the first flight to take humans all the way to the Moon, a huge step forward in developing a lunar landing capability. The mission lasted 7 days and included 10 orbits around the Moon. It achieved many other firsts: It was the first manned spacecraft to travel atop the powerful Saturn V booster, the first crewed launch from NASA's new Kennedy Space Center, and the first time in history that humans saw the lunar surface up close and with their own eyes. As the world watched on their televisions sets on Christmas Eve, December 24, 1968, the crew of Apollo 8 read verses from the Book of Genesis as they orbited the Moon.

Apollo 8 provided us with the first picture ever taken by a human of the Earth from deep space, a view that poignantly showed how we are a single part of a much larger system. That now iconic snapshot of Earth peeking out from beyond the lunar surface as the first crewed spacecraft circumnavigated the Moon has been called "Earthrise." Some credit that snapshot of Earth with inspiring the environmental movement initiated on the first Earth Day in 1970.

To test the Apollo lunar landing module for its future assignment to carry humans to the surface of the Moon, Apollo 9 lifted off in March 1969. Over the mission's 10 days in Earth orbit, the crew appraised the lunar landing craft as a self-sufficient spacecraft. They also performed active rendezvous and docking maneuvers paralleling those scheduled for the following Apollo 10 lunar orbit mission. The goal of this Apollo mission, successfully achieved, was to simulate maneuvers near Earth that would be executed in actual lunar missions.

Just two months later, NASA put the collected wisdom of learned lessons together to run through all aspects of a human lunar landing, minus the actual lunar touchdown. In May 1969, Apollo 10 astronauts lifted off to operate around the Moon—a complete staging of the Apollo 11 mission that would follow. The fourth crewed mission in the United States Apollo program, this dress rehearsal for a Moon landing brought the lunar module as close as 47,000 feet from the lunar surface.

Whereas the Apollo 10 mission provided key data needed to green-light a lunar landing, this flight was not without incident. Following the lunar lander's descent stage separation and ascent engine ignition, the craft started to roll violently, causing the crew to utter expletives as they struggled to regain control of their vehicle. It was a harrowing eight seconds brought about by a malfunction in the backup guidance system. The astronauts solved the problem and faced no further difficulties in the mission.

Now it was time for the real thing—the first attempt to land humans on the surface of the Moon.

On July 16, 1969, at 9:32 a.m. EDT, a Saturn V rocket lifted off from its pad at NASA's Kennedy Space Center, launching Apollo 11 on its way to the Moon. All of the tests, all the technology development, the robotic imagery, trial runs in Earth orbit and at the Moon—all led to that moment. Atop the mammoth booster on its way to the Moon were three American men who would make their marks in the history books. Hundreds of thousands of spectators gathered to celebrate the send-off, a nation-backed undertaking by the United States on its way to successfully beating the Soviet Union in the race to put people on the Moon. Roaring into Florida skies, the Saturn V engines strained against Earth's gravity, rattled windows miles away, and hurled the Apollo 11 crew on a route to the Moon . . . and on a course to fulfilling President Kennedy's bold goal.

CHAPTER FOUR

MAN ON THE MOON: APOLLO'S LASTING LEGACY

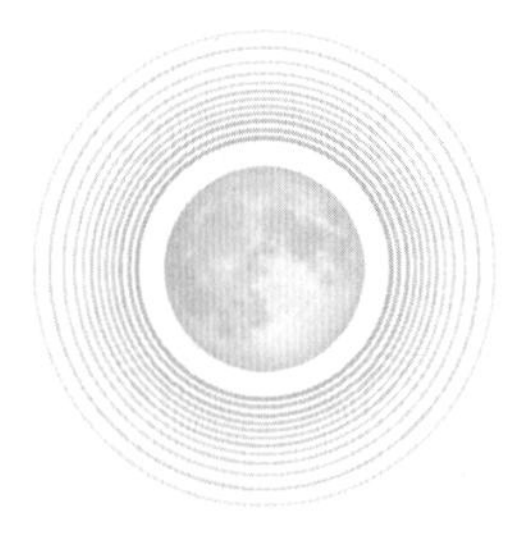

FIFTY YEARS AGO—a half century in time and space—two astronauts carefully descended a nine-rung ladder attached to their fragile spaceship, called the *Eagle*. Each took one small step to set foot on the time-weathered, dusty, and desolate Moon. The voyage from Earth was the culmination of a bold, risky gambit—shouldered by some 400,000 people who banded together to make real a vision. It was a unified enterprise, a mix of government, industry, and academic wherewithal that transformed a long-held dream into reality.

On July 20th, 1969, Neil Armstrong and Buzz Aldrin became the first humans to walk on the Moon. Apollo 11 colleague, Mike Collins, remained in lunar orbit to receive the Moon visitors returning after their brief but history-making stopover on the uninviting lunar land. He watched from the command module as the *Eagle* began its traverse to the Moon's terrain and would later recall that the lunar lander was the weirdest looking contraption he had ever seen in the sky.

During the *Eagle*'s final minutes of descent to the lunar surface, its navigation and guidance computer distracted the crew with the first of several unexpected "1202" and "1201" program alarms. The computer was trying to do too many things at once, Aldrin later pointed out. "Unfortunately," he said, "it came up when we did not want to be trying to solve these particular problems."

As the *Eagle* descended into the Moon's Sea of Tranquility, Armstrong had to improvise, manually piloting the ship past an area filled with rocks. The onboard computer had guided them to land on the side of a large crater with steep slopes littered with huge boulders. "Not a good place to land at all," Armstrong later said. "I took it over manually and flew it like a helicopter out to the west direction, took it to a smoother area without so many rocks, and found a level area and was able to get it down there before we ran out of fuel. There was something like 20 seconds of fuel left," he explained in a 2012 interview.

Armstrong confirmed that landing was his biggest concern: The unknowns were rampant, and there were a thousand things to be worried about. "I thought we had a 90 percent chance of getting back safely to Earth on that flight but only a 50-50 chance of making a landing on that first attempt. There are so many unknowns on that descent from lunar orbit down to the surface that had not been demonstrated yet by testing, and there was a big chance that there was something in there we didn't understand properly and we had to abort and come back to Earth without landing."

After the lunar module settled onto the Moon, Armstrong radioed: "Houston, Tranquility base here. The *Eagle* has landed." Capsule Communicator and Apollo astronaut Charles Duke responded: "Tranquility, we copy you on the ground. You got a bunch of guys about to turn blue. We're breathing again. Thanks a lot."

More than six hours later, Armstrong departed the *Eagle* and reached the foot of the ladder, observing that the landing craft footpads were depressed in the surface about one or two inches. This observation was followed by his stepping onto the Moon. Joined by Aldrin, both did not experience any difficulty in adapting to one-sixth gravity. The combination of space suit and backpack that sustained their life on the lunar surface operated superbly. "The primary difficulty was just far too little time to do the variety of things we would have liked. We had the problem of the five-year-old boy in a candy store," Armstrong later said, quoted in NASA's *Apollo Expeditions to the Moon,* published in 1975.

About seven minutes after stepping onto the Moon's surface, Armstrong snagged a soil sample, using a sample bag on a stick and grabbing the lunar specimen early to guarantee there would be some lunar soil brought back in the event that an emergency required the astronauts to ditch their moonwalking and scurry back into the *Eagle.* He tucked that bag into a pocket on his right thigh. Moving in the lunar gravity was "even perhaps easier than the simulations," recalled Armstrong. "It's absolutely no trouble to walk around."

"My first words as I took foot on the lunar surface," Aldrin said, "were 'magnificent desolation.' It was a 'magnificent' accomplishment for man to set foot on another world for the first time. And yet there was the 'desolation' of the lunar landscape with no sign of life, no atmosphere and total blackness beyond the sunlit terrain."

Aldrin tested methods for moving around, including two-footed kangaroo hops. "I took off jogging to test my maneuverability," Aldrin recounted. "I noticed immediately that my inertia seemed much greater. Earth-bound, I would have stopped my run in just one step, but I had to use three of four steps to sort of wind down. My Earth weight, with the big backpack and heavy suit, was 360 pounds. On the Moon I weighed only 60 pounds."

After spending approximately two hours and 31 minutes on the surface, the astronauts ended their moonwalks, returned to the lander, got some sleep, and then rocketed off the Moon and successfully reconnected with astronaut Michael Collins in the command module. Collins recalled their momentous reunion. "The first one through is Buzz, with a big smile on his face," he said. "I grab his head, a hand on each temple, and am about to give him a smooch on the forehead, as a parent might greet an errant child; but then, embarrassed, I think better of it and grab his hand, and then Neil's. We cavort about a little bit, all smiles and giggles over our success . . . and then it's back to work as usual," he recounted in *Apollo Expeditions to the Moon*.

"Even though we were all farther away than humans had ever been, we always felt connected to home," Aldrin has since remembered. That feeling seemed mutual. Back on Earth, an estimated 600 million people witnessed on television the momentous Apollo 11 landing. Ninety-four percent of American televisions tuned in to view the historic event, which took place just before 11 p.m. eastern daylight time. That fact made President John F. Kennedy's proclamation eight years earlier even more prophetic: "It will not be one man going to the Moon, it will be an entire nation."

The voyage of Apollo 11 was a product of nearly a decade of work, hardships, tragedy, and awe. It all began earlier, with a plan. President John F. Kennedy had made clear in September

1962 what the perilous undertaking would involve. Buzz Aldrin keeps in mind why the world welcomed the crew of Apollo 11: "We understood that people were not cheering for three guys, but for what *we* represented that by the world coming together, *we* had accomplished the impossible." During the crew's celebratory jaunt around the world, Aldrin remembers seeing signs hoisted in the air that said, "*We* Did It!"

"The true value of Apollo is the amazing story of innovation and teamwork that overcame many obstacles to reach the Moon," Aldrin says. His words are echoed by the late Gene Cernan, Apollo 17's commander. As he departed the Moon's surface he spoke of the legacy of Apollo balanced with a future yet to transpire: "As I take man's last step from the surface, back home for some time to come . . . but we believe not too long into the future . . . I'd like to just say what I believe history will record: That America's challenge of today has forged man's destiny of tomorrow. And, as we leave the Moon at Taurus-Littrow, we leave as we came and, God willing, as we shall return: with peace and hope for all mankind."

—||—

SIX APOLLO LUNAR LANDING MISSIONS took place between mid-1969 and the end of 1972. From the Apollo 11 moonwalkers, who spent two and a half hours high-stepping across the terra incognita, to Apollo 17's 22 hours of surface forays, the stopwatch for humans setting foot or wheels on the Moon tallies to 80 hours, less than four days total. But in those hours, the 12 moonwalkers left the footprints that we still vividly picture and brought home samples that would continue to reveal discoveries to scientists.

During the next mission to the Moon, Apollo 12 in November 1969, the crew spent 31.5 hours total on the lunar surface, with 7.5 hours during two moonwalking periods. The mission's *Intrepid* lunar module made a precision landing on the lunar surface, shoring up confidence that future exploration landing points could be targeted to rougher terrain and specific areas of scientific interest. The crew landed so precisely, in fact, they were within walking distance—600 feet—of the Surveyor 3 spacecraft that had set down on the Moon in April 1967. They photographed the robotic lander, retrieving equipment from the probe, including its scoop and television camera. They collected rock samples as well, mostly basalts, dark-colored igneous rocks hundreds of millions of years younger than the rocks collected on Apollo 11.

The Apollo scorecard of piloted lunar landings was marred in April 1970 when Apollo 13 did not attain a Moon touchdown. The flight of Apollo 13 would have been a calamity and a tragedy if it were not for the resolve of the astronauts and the perseverance of mission controllers. A crippling failure en route to the Moon curtailed the flight of Mission Commander James Lovell, Command Module Pilot John "Jack" Swigert, Jr., and Lunar Module Pilot Fred Haise. In transit to the Moon, Lovell later recalled, he looked out the window and saw oxygen escaping from the spacecraft's rear. He and his crewmates were in serious trouble. "Houston, we have a problem," he famously reported.

Tackling the serious threats caused by the vessel's inadequate power levels, loss of cabin heat, and the critical need to cobble together a carbon dioxide removal system, the crew returned safely to Earth, splashing down in the South Pacific Ocean. "The flight was a failure in its initial mission," Lovell said, but he granted that it was "a tremendous success in the ability of people to get together, like the mission control team working with what they had and working with the flight crew to turn what was almost a certain catastrophe into a successful recovery."

The Apollo program came back on line with Apollo 14, which landed in February 1971, just 87 feet from its targeted landing point. The astronauts used a modularized equipment transporter (MET) to haul equipment during two moonwalking periods. Fondly known as the "rickshaw," the two-wheeled MET was designed as an aid to surface exploration. Serving as a workbench with a place for lunar hand tools, it also carried cameras, sample containers, spare film, and a portable magnetometer. During two traverses totaling more than nine hours on the lunar surface, the astronauts collected 94 pounds of rocks and soil for return to Earth. The crew's moonwalks covered a total distance of a little over two miles and involved sample collecting at 13 locations, deploying or performing 10 experiments, and examining and photographing the lunar surface.

In a brief moment of lunar levity, Commander Alan Shepard pulled out a makeshift six-iron club and sent two golf balls flying across the lunar surface. Shepard's thick gloves and rigid space suit forced him to swing the club with one hand. He viewed his prank as somewhat scientifically valuable. With little atmosphere and much lower gravity, golf balls on the Moon should travel much farther than on the Earth. And with all those craters, a hole in one should be easy!

The following three Apollo lunar landing missions each increased in duration and benefited from the use of the two-seat lunar roving vehicle to haul scientific instruments to various sites and return Moon samples to their landing craft. Deployed onto the lunar surface during the Apollo 15 mission, the lunar rover extended the distance astronauts could travel and expanded the tasks they could accomplish during their limited surface time.

Apollo 15 was the first in a series of three advanced missions during which crews explored the Moon over longer periods and over greater ranges with more instruments for collecting scientific data than on previous missions. At the end of the last Apollo 15 moonwalk, Commander David Scott performed a live demonstration, holding out a heavy geologic hammer and a feather, dropping them at the same time. The objects underwent the same acceleration and struck the lunar surface simultaneously. In a vacuum, there is no air resistance, and the hammer and feather fell at the same rate—a test of Galileo's law of motion concerning falling bodies, lunar style.

During less than three days of stay time, and thanks to their lunar rover, Apollo 16 astronauts collected samples from 11 sites, including a drill core from seven feet below the Moon's surface. The major objective of the mission was to investigate a region and type of terrain that had not previously been visited, the central lunar highlands: elevated areas riddled with craters and rocks formed by the intense bombardment the lunar surface endured in the early days of its existence.

The last Apollo mission reached the Moon on December 11, 1972. Apollo 17 astronauts stayed on the lunar surface for 75 hours and traveled about 22 miles. Scientific objectives for the mission included geological surveying, materials sampling, and surface feature observation in the area of the Taurus-Littrow region on the eastern rim of Mare Serenitatis, with the goal in mind of investigating the possibility of young, explosive volcanism in this region. One of the astronauts on this mission was geologist Harrison "Jack" Schmitt, the only professional scientist to have visited the Moon. Keenly involved in training previous Apollo astronauts, now Schmitt could survey the lunar surface in person.

When the Apollo 17 threesome returned home on December 11, 1972, the first phase of human exploration of the Moon ended. That was more than 46 years ago.

—||—

ALTHOUGH THE UNITED STATES enjoyed its victorious Apollo program, the Soviet Union struggled with their lunar projects. The robotic Luna 15 launched just three days before liftoff of the historic Apollo 11 mission to the Moon. Soviet space engineers attempted for a second time to robotically scoop up and bring lunar specimens back to Earth. Luna 15's goal was unquestionably to upstage the first human Moon landing. The Soviet craft did indeed reach lunar orbit. As Apollo 11's Armstrong and Aldrin walked on the Moon, Luna 15 fired its main retrorocket engine to initiate descent to the lunar surface. However, mission control lost contact with the robot, and Luna 15 likely struck a mountain during its trajectory to land on the Moon.

Ceding the fact that the first humans on the Moon were born in the United States was not easy for the Soviet Union. In the end, the Soviets abandoned their humans-to-the-Moon program, made disastrously apparent after four uncrewed flights failed. Their N1 super heavy-lift booster, the Soviet equivalent of the American Saturn V moon rocket, made its last takeoff in November 1972. The flight catastrophically ended 107 seconds later—the megabooster's longest flight up to that moment—but brought to an end any hope of placing cosmonaut footprints on the Moon.

THE 2015 NASA REPORT on the science training of the Apollo astronauts, written by William C. Phinney, explains the many lessons learned from the Apollo expeditions. Their lunar surface activities fell into two broad categories: exploration and operation. Lunar visitors needed to be prepared for the challenges and opportunities their short time on the Moon's surface offered, and they needed to be completely versed in the operational and procedural context necessary for those explorations. During the six successful Apollo landing missions, a great deal was learned about living and working in the lunar environment—lessons that future Moon mission planners will likely incorporate when planning future treks.

Planners asked what crews should do on the Moon after they stepped onto the barren landscape, and once the decision was made that science activities, particularly geoscience, should be pursued, considerable debate ensued over how to achieve scientific tasks. Mission planners had to decide what instruments and tools to send to the Moon, what samples and photos were to be returned, which landing sites served the best purpose, and the most effective makeup of the crews. Crewmembers had to be trained to recognize and evaluate key features of landforms and samples and then use that knowledge in deciding how to best to use their available time.

Mission planners plotted out traverses on the Moon. The astronauts were trained to use tools, set up instruments, take photos, collect and document samples, and provide proper descriptions. Their colleagues in lunar orbit were taught how to manage additional instruments and take photos from lunar orbit. Over a year, each three-person crew averaged nearly 100 hours a month in training for these activities.

Operational training required a familiarity with procedures and equipment to accomplish work on the Moon efficiently , with problems handled on-site or discussed knowledgably with support personnel back on Earth. Most of the Apollo moonwalkers were well prepared for both explorations and operations.

But still, problems cropped up as astronaut training progressed. Working with mission planners, they had to revise awkward procedures, determine the best means of communications between all involved groups, and devise contingency actions for any real-time problems that arose on the Moon. Then there were personal obstacles: adjusting groups of personnel for the best combination of backgrounds, personalities, and skills.

After all that training, perhaps Apollo's true lasting legacy is the science the program was able to achieve. The astronauts brought back Moon samples that have become the gift that keeps on giving. Owing to advances in analysis techniques and laboratory equipment, researchers today continue to tease out even more information from the 842 pounds of lunar rocks, core samples, pebbles, sand, and dust brought to Earth from the lunar surface.

Some operational hazards can be practiced; others simply must be planned for, as the Apollo program demonstrated. Using the Moon as a training ground for living "off planet," planners and astronauts must tackle the issue of radiation—a problem for those on the surface, those orbiting the Moon, and even those traveling there. Radiation will remain a problem for any human living away from Earth, in any space environment. Looking back, we now recognize that the lunar module that took Apollo crewmembers to the Moon's terrain did not provide much shielding. During moonwalks, crews were vulnerable—but they were lucky: One of the strongest solar flares on record happened in August 1972, between the Apollo 16 and 17 missions, but no such burst of radiation happened when anyone was actually on the Moon.

Apollo astronauts still experienced high-energy radiation as they traveled to the Moon and back. Cosmic ray visual phenomena, or "light flashes," are spontaneous flashes of light visually perceived by some astronauts outside the magnetosphere of the Earth. Crewmembers of the Apollo 11 mission were the first astronauts to describe this visual phenomenon. During their trans-Earth coast, both Armstrong and Aldrin reported seeing faint spots or flashes of light when the cabin was dark and after their eyes had become dark-adapted. They reported seeing the light flashes with eyes either open or closed, suggesting that the flash effect is produced by cosmic radiation penetrating the optical nervous system.

The high-radiation environment of outer space is a major constraint for long-duration operations on the lunar surface, a challenge future missions need to plan for more rigorously if their work includes longer time on or around the Moon. Crews will need shielded habitats to protect against exposure to the cosmic ray background over stays of significant lengths. They will also need emergency sheltering during forays away from permanent facilities in case they encounter solar flares, infrequent but dangerous and unpredictable, which spew high doses of high-energy particles.

Another major operational hazard that future Moon missions need to tackle head-on is lunar dust. It gets everywhere. The lunar surface is laden with tiny, sharp-edged particles that intrude into everything, from space suit zippers to moving parts of equipment, cameras, and anything else you can think of. Astronauts can track dust into their landers or habitats, where it could get into the cabin atmosphere and become an irritant to the crew's respiratory tracts. They experienced minor irritations and possible hay fever symptoms from ingestion of lunar dust particles. After doffing their helmets and gloves, the astronauts could feel the abrasive nature of the dust coating their skin, as well as smell and even taste the Moon.

The lunar dust, or regolith, contains several types of reactive dust, including silicon dioxide (50 percent), iron oxide and calcium oxide (4 percent), and other oxides. Silicon dioxide is highly toxic; dusts containing silica on Earth are responsible for silicosis, a life-threatening lung disease found mainly in stonemasons. Dust particles lodge in different areas of the lungs, depending on the particle size; nanoparticles penetrate the deepest. Lower gravity influences the effects of nanoparticles inside the body, which means the ensuing health effects of exposure are different on the Moon.

A consistent and persistent message from the Apollo moonwalkers: The Moon is a Disneyland of dust. "I think dust is probably one of our greatest inhibitors to a nominal operation on the Moon," said Apollo 17 moonwalker Gene Cernan during a post-flight technical debriefing. "I think we can overcome other physiological or physical or mechanical problems—except dust."

Future long-duration missions need to find a way to manage the dust problem. One possible solution is to collect and direct solar energy to melt the nearby lunar topside into a clean, glassy apron, reducing the chance of contamination. To protect vehicle and habitat interiors, perhaps astronauts can dock their the suits outside of the lander and only rarely bring them inside for maintenance. Crews will need to clean gear more thoroughly before bringing it inside than was possible during Apollo missions.

Another challenge that must be met is to improve astronauts' ability to manipulate tools and the environment. The Apollo gloves imposed serious limitations on hand mobility, finger dexterity, and tactility. Their cumbersome thickness and weight resulted in serious forearm fatigue, which began within minutes of the start of any extravehicular activity and continued all day without letup. Try as they might, Apollo astronauts experienced damage to their fingers and fingernails arising from their gloves. Work is already under way to develop more flexible suits and better gloves than those used in the Apollo program.

—||—

DESPITE ALL THESE PROBLEMS, the scientific bounty that came from the Apollo missions is tremendous. From 1969 deep into 1972, Apollo moonwalkers dashing back to Earth transported a total of nine containers of lunar material, sealed on the lunar surface and brought back to Earth for detailed examination. Three sealed samples, one each from Apollo 15, 16, and 17, remain unopened, intentionally saved until technology and instrumentation has advanced to the point that investigators can maximize the scientific return on these unique specimens. Some leading lunar researchers say that now the time is right to open at least one of the conserved sample containers. With renewed interest in the Moon, further knowledge about the volatiles in the lunar soil—such as water, hydrogen, and methane—will be useful for planning future human involvement with the Moon and its materials and in the design of future lunar missions.

The Lunar Sample Laboratory Facility, a special building at NASA's Johnson Space Center in Houston, houses the geological samples from the Apollo missions. They are physically protected, environmentally preserved, and scientifically processed. From six different exploration sites on the Moon, the Apollo collection consists of 2,200 separate lunar specimens. Scientists have used some of those samples to learn that the

Moon's interior has water. A sensitive mass spectrometric analysis of basaltic glasses brought back to Earth by the Apollo 15 and Apollo 17 missions led to the discovery.

Laboratory analyses of glass and beads collected from the Moon's surface indicate that, unlike most lunar materials, they have volatile element–enriched coatings from their gas-rich source regions. That is, they originated at great depths inside the Moon, and they represent the most basic or primitive of lunar volcanic materials. Studies of the lunar pyroclastic materials thus provide unique information on the characteristics of the lunar interior and the origin, evolution, and early differentiation of the Moon.

Although Apollo astronauts brought lunar material home, they also left scientific experiments up and running on the Moon's surface. Consider it an even trade. Apollo 11, 14, and 15 each deployed reflectors for the ongoing Lunar Laser Ranging experiment, which uses short-pulse lasers and state-of-the-art optical receivers and timing electronics to receive signals from Earth. Specially outfitted Earth stations send out intense laser shots, like little "bullets" of light, which hit the arrays on the Moon and bounce back. Earth stations record the time it takes for the laser bursts to return. It can take anywhere from 2.34 to 2.71 seconds, depending on how far away from Earth the Moon is at the time. (The value isn't constant because the Moon's orbit around Earth isn't a perfect circle.) Laser ranging has produced many important measurements, including an improved knowledge of the Moon's orbit. The experiment also lets researchers measure the rate at which the Moon is receding from Earth, calculated at currently 3.8 centimeters a year.

It also has shown deviations in the Moon's rotation, which are related to the distribution of mass inside the Moon; the variations imply the existence of a small core.

The Apollo reflectors are marvels of engineering, designed and built on an ambitious six-month time frame. Great attention went into thermal considerations so that the Moon-situated reflectors might perform well in both dark and sunlight conditions, although lunar dust thwarts their careful design. Despite age, degradation, and the fact that they were designed for 1960s-era laser technology, they continue to this day to deliver important scientific results.

ONE OTHER THING THAT WAS LEFT on the Moon was the American flag—an exceptionally powerful symbol of American identity and national pride. When Apollo 11 astronauts Armstrong and Aldrin planted the U.S. flag just inches into the lunar soil (as deep as they could due to the harder-than-expected surface below the top powdery layer), it was strictly a symbolic activity. "Nevertheless, there were domestic and international debates over the appropriateness of the event," explains Anne Platoff, a former NASA contractor who led a report in the early 1990s about the political and technical aspects of placing the American flag on the Moon.

Initially composed in 1962 and ratified in 1967 by an assemblage of the day's spacefaring nations, a United Nations Outer Space Treaty precluded any territorial claim on the Moon or any other planetary body. Outer space, moons, and planets are free for all to explore, and no sovereign claim can be made to any of them, the treaty states. The United States, as a signatory to the UN treaty, could not claim the Moon. Raising a flag on the lunar surface could merely be an emblematic gesture, an expression of triumph similar to the planting of a flag on Mount Everest or at the North and South Poles. The legal status of the Moon clearly would not be affected by the presence of the American flag on the lunar surface, and NASA was well aware of the international controversy that raising such a flag might incur.

To deal with the hail-to-the-flag hullabaloo, a Committee on Symbolic Activities for the First Lunar Landing was formed. The committee was instructed to choose symbolic actions that would not jeopardize crew safety or interfere with mission objectives. Activities selected would signal the first lunar landing as a historic forward step for all mankind, that though accomplished by the United States, would not give the impression that America was taking possession of the Moon. Flag ceremonies on the Moon were planned in keeping with President Kennedy's declarations: ". . . we mean to lead [the exploration of space], for the eyes of the world now look into space, to the Moon and to the planets beyond, and we have vowed that we shall not see it governed by a hostile flag of conquest, but by a banner of freedom and peace."

In fulfilling their charge, the committee pondered several options. In the end, they decided to hoist the flag of only the United States during the moonwalk but to leave behind a plaque bearing an inscription emphasizing that the reason for the mission was exploration and not conquest: "Here men from the planet Earth first set foot upon the Moon July 1969, A.D.

We came in peace for all mankind." It featured pictures of the Eastern and Western Hemispheres of Earth to symbolize the crew's point of origin.

The famous Apollo 11 flag raising ceremony on the ancient lunar surface took all of 10 minutes during Armstrong's and Aldrin's two-and-a-half-hour moonwalking venture, but no doubt for many back on Earth, it was one of the mission's most memorable aspects—as reflected in the controversy over its exclusion from the recent film on Apollo 11, *First Man*. At the time, there were no formal protests from other nations, no angry TV viewers: No one seemed to see the flag raising as an illegal attempt to claim the Moon. In addition to the large three-by-five-foot U.S. flag left behind on the Moon, four-by-six-inch flags of the 50 states, the District of Columbia, the U.S. territories, all member countries of the United Nations, and several other nations were carried inside the lunar module and returned for presentation to governors and heads of state after the flight returned.

Each subsequent Apollo crew planted an American flag on the lunar surface, three different sizes. The Apollo 17 flag was noteworthy as it had been located in the Mission Operations Control Room during the other Apollo missions and then placed on the Moon by the last lunar landing crew.

Raising a flag on the Moon presented NASA engineers with an interesting technical challenge. Without an atmosphere for any wind to flow through, any flag placed on the Moon would hang limp. To make each flag appear to fly, engineers designed a flagpole with a horizontal bar along the top. And they couldn't be just any flags: "Weight, heat resistance, and ease of assembly" by an astronaut in a space suit also had to be taken into account in the design of the Moon-destined flags, former NASA contractor Platoff explains.

Now, decades later, no one knows the condition of those flags. Could they have even remained standing, given the blast from lunar module ascent engines as crews rocketed away from the Moon? Almost certainly they are not in the same condition as when they were first hoisted on the lunar surface. Prolonged exposure to sunlight likely has made the flags brittle. Perhaps they have disintegrated over time. Meteoroids could have damaged them. Only new missions can reveal whether the flags have withstood the decades of the Moon's environmental conditions.

THE APOLLO MISSIONS left behind more than scientific instruments and American flags. They created a legacy, and to retain that legacy there is need to safeguard that history.

The spot where Apollo 11's *Eagle* lander touched down in the Sea of Tranquility on July 20, 1969, is the most extraordinary archaeological site in history, says Beth O'Leary, a leading authority in the anthropology of space exploration at New Mexico State University. Imagine if, here on Earth, gone were all traces of humankind's historical past—whether Stonehenge in England, the pyramids in Egypt, or the 2.5-million-year-old stone tools found in Africa's Olduvai Gorge. "Only preservation and documentation of the objects and events allows humans to accurately reconstruct and remember the past," says O'Leary, who has been guiding the "Lunar Legacy Project" in an effort to study and catalog the Apollo leftovers from the six historic but brief stays on the Moon. A total of 190 tons of cultural material was left on the Moon, O'Leary estimates, and about 106 artifacts are specific to the Apollo 11 site.

The Apollo 11 moonwalkers collected lunar samples, deployed experiments, and took in the sights as they snapped photographs of their eerie surroundings. They placed on the surface several memorials: a mission patch to memorialize the tragic launchpad fire in 1967 that took the lives of Apollo 1 astronauts Gus Grissom, Edward White, and Roger Chaffee; and a satchel holding medals commemorating two deceased Soviet cosmonauts, Yuri Gagarin and Vladimir Komarov. They left behind symbolic artifacts like a gold olive branch, the traditional symbol of peace.

But they also littered the lunar surface. By the end of their two and a half hours of scurrying around, they discarded prosaic items like astronaut boot coverings, portable life support backpacks, food wrappers, and a hammer. They discarded junk, like plastic coverings, insulation material, brackets, a TV camera, tongs and scoops, even urine and defecation collection devices. Well over a hundred items were discarded on the Moon. In short, littering was planned as a way of lightening the load of the lunar blastoff back to the orbiting spacecraft. And of course, there are those momentous boot prints, the trails Armstrong and Aldrin made as they reconnoitered the landing region. O'Leary sees the entire site and all of its pieces as an archaeological legacy. Although Apollo 11 astronauts knew they were taking a giant leap for humanity, they likely did not take into account that they were creating a lunar legacy, one that must be preserved for future generations.

O'Leary and like-minded Lisa Westwood, a professional archaeologist in Sacramento, California, and director of Cultural Resources at ECORP Consulting, Inc., have been dogged in their campaign to name Tranquility Base a historic landmark. Once a property is designated a historic landmark, it can be nominated for inclusion on the UNESCO World Heritage List. Westwood and O'Leary also have been working with the International Council on Monuments and Sites (ICOMOS), the World Heritage List's advisory body, over the last several years to make the case for the Apollo sites to be added to this list. Others have joined in to seek international protection of the Apollo lunar landing sites. For All Moonkind, Inc., is a nonprofit group that has orchestrated a multipart plan to work with the United Nations and the international community to preserve each of the six lunar landing sites as part of our heritage. The strategy outlined encompasses use of an international team of space lawyers, policymakers, and marketers to achieve the organization's goals. For All Moonkind wants to be able to deliver a formal plan—already vetted by national space agencies—to the United Nations Committee on the Peaceful Uses of Outer Space.

Politicians are also concerned about protecting the landing sites. In March 2018, the White House Office of Science and Technology Policy (OSTP) issued a report on "Protecting and Preserving Apollo Program Lunar Landing Sites and Artifacts." In the report, it reviews the status of the six sites where Apollo crewmen touched down on the Moon, the benefits of preserving the artifacts left on the lunar surface, and the existing legal foundations that provide for the sites to be left undisturbed. The report recognizes that international and commercial plans to send robotic spacecraft and humans to the Moon could cause a significant amount of damage to the earlier sites, if they land on top of or too close to them. Rocket exhaust plumes, for example, might blast away the celebrated footprints and rover tracks.

"Risks to damage lunar heritage sites must be balanced against other national and international interests," the OSTP report reads. The lunar heritage sites can be protected at a reasonable cost, the assessment continues, while still fostering commercial space activities and government-sponsored missions back to the Moon. In its report, the White House OSTP did raise a concern about hindering efforts by U.S. companies to send commercial robotic missions to the Moon. At least one private company planning to go to the Moon is specifically targeting the vicinity of an Apollo landing site for its first mission.

There may be scientific and engineering value in revisiting and perhaps even removing and returning hardware to Earth from some of the other sites at some point in the future. Those plans need to incorporate the concerns of historians, archaeologists, and others about the cultural and historical nature of the locations. Arguably the best solution could be a museum on the Moon. Tourists could see firsthand how the human exploration of our neighboring worlds began. A first step is preserving the Apollo sites now, so that the opportunity still exists in the future.

—||—

WE CAN VALUE THE LEGACY OF APOLLO in other ways. The reverberations from 20th-century Project Apollo continue to waft through our 21st-century society—but in a way that heralds striving for goals once thought unobtainable. How often do you hear, "If we can go to the Moon, why can't we _______?" (You fill in the blank.) We speak nowadays of quests as being "moonshots," a word perhaps anchored in an earlier turn of phrase, "shoot for the Moon." Collectively, these Apolloesque idioms imply aspiring for a lofty goal, and nowadays we recognize that we can do it.

British science fiction writer and futurist Arthur C. Clarke wrote that every revolutionary idea passes through three stages:

1. It's completely impossible.
2. It's possible, but it's not worth doing.
3. I said it was a good idea all along.

Today we use the word "moonshot" to connote an ambitious, exploratory, and groundbreaking project. Google X's "Moonshot Factory" has embraced the term for inventive initiatives like driverless cars, robots for manufacturing purposes, even life extension. They define "moonshot" as a task or idea that addresses a huge problem, proposes a radical solution, and makes use of breakthrough technology. The National Cancer Institute at the National Institutes of Health is sponsoring the "Cancer Moonshot" to accelerate cancer study and make more therapies available to more patients, while also advancing the capacity to prevent cancer and detect it at an early stage. In 2018 the President's National Security Telecommunications Advisory Committee reported on a "Cybersecurity Moonshot," noting it would be "a societal transformation rather than a singular, visual triumph."

The true moonshot, the fulfillment of Project Apollo, represented for us all a bridge crossing from science fiction into science reality and the audacious success of humans now able to boldly go where no others have gone before.

ONE OF THE WAYS WE CAN APPRECIATE the full extent of the Apollo program and its legacy is through the Apollo Lunar Surface Journal, an online record of operations conducted by the six pairs of astronauts who landed on the Moon from 1969 through 1972, and created by enthusiasts Eric Jones and Ken Glover. It is a treasure trove of details. Perusing its pages reveals many insights. The journal makes clear that beyond the symbolism of Apollo 11's touchdown, the entire program was above all a step-by-step learning experience. From mission to mission, the astronauts gained self-assurance in themselves and their equipment, learning with each effort how to get good work done more proficiently. Except for the hours the astronauts spent resting and trying to sleep, they were constantly working, trying to maximize the return from these all-too-brief opportunities for exploration. In all, the Apollo crews spent about 24 man-days on the Moon and, although some activities were repeated from day to day and from mission to mission, the learning curve was steep and the frequency of novel experiences was high.

Never in the history of exploration has there been an undertaking quite like Apollo. Changes in human perspective usually come in small steps. But Apollo was a rare and extraordinary opportunity to make a purposeful leap. It is evident that for those working on the program, Apollo was more than a race to beat the Soviet Union; it was the start of an even grander enterprise. Some compared the first Moon landings with the first steps by sea creatures onto the land, and few in NASA believed that Apollo was an end in itself. As with the ancient colonization of the dry land by creatures from the sea, there is more to come. The long-term potential of the Moon and the rest of the solar system will only be realized after we learn how to live and work in new off-Earth environments—and Apollo was a start.

Jump to thoughts of a future lunar base. The knowledge the Apollo astronauts gained has much to tell us: about the challenges ahead of working on the Moon, and about taking advantage of lunar conditions to make work easier in the weak gravity field and safer in the high-radiation situation. The Apollo experience will matter to engineers and planners trying to design equipment and procedures. More to the point, as the journal entries underscore, they will need to know what worked, what didn't work, and why. "And they would do well to pick up where Apollo left off," reflects the journal's founder and editor emeritus, Eric Jones.

—||—

THE HUMAN OUTREACH to the Moon was a massive engineering challenge. Leading space officials of the time saw Apollo's real purpose as advancing technology, but the effort unquestionably demanded heaps of political will and money. The Apollo program cost $25 billion, many times higher than today's conversion equivalent. How did space exploration, once the purview of rocket scientists, rally public attention and support? Likewise, what was behind a government program's ability to turn its greatest achievement into a communal experience—and can we do that again?

Americans were an enthusiastic audience for NASA's pioneering "brand journalism," primed by science fiction, magazine articles, and appearances by rocket and space visionary Wernher von Braun on the "Tomorrowland" segments of the *Disney* prime-time television show. NASA and its many contractors made sophisticated efforts to market the facts about space travel. American astronauts signed exclusive agreements with *Life* magazine, making them the heroic and flag-waving faces of the program. Mix in shrewd product placement: Hasselblad was the "first camera on the Moon"; Sony cassette recorders and supplies of Tang were touted as onboard spacecraft; and astronauts used the Exer-Genie personal exerciser. Everybody wanted a place on the space bandwagon.

Richard Jurek, a public relations expert and authority on how the Moon was marketed in the space race era of the 1960s, explains that the chief lesson from Apollo was the importance of television as the medium that engaged the public and brought them along on the adventure. The merit of Moon exploration for sheer science is a hard sell to the general public, requiring a different skill set and approach. Put bluntly, people want to know what is in it for them, which demands promoting the fascination while also selling the scientific aha moments.

PROJECT APOLLO IS NOW in the rearview mirror of history. What is the best way to embark on a new, goal-driven space exploration trajectory? A space effort comparable to Apollo can draw upon today's technology and knowledge, leading to a Moon-Mars initiative—a much larger program and a multi-decade effort to get first to the Moon and then to Mars. This project needs the unequivocal and sustained commitment of the nation—indeed, many nations and private entities—even more so than was required for the Apollo program.

Apollo was a success due to a sufficient base of technology and a large reservoir of young engineers and skilled workers. It also succeeded thanks to an articulate, persuasive, and devoted president and a committed Congress willing to spend dollars to assure that unanticipated problems could be resolved. And of course, it all happened within the environment of pervasive national unease, stemming first from Sputnik and then from the catalytic event of the Soviet Union's flight of cosmonaut Yuri Gagarin in April 1961, the first human to orbit the Earth.

So what do we need to further a new U.S. space agenda today, especially a Moon-Mars deep-space program? We need improved education in science, technology, engineering, and math (so-called STEM) skills, along with critical thinking. We need to maintain an average workforce age of less than 30 years.

And then there's the risk factor: Maybe we need to eliminate the political aversion to taking necessary risks.

As the Apollo Moon landing missions progressed and became more science-oriented, attitudes changed. It wasn't just about getting to the Moon and being the first. Now, it was about understanding the place. Astronauts were well aware that doing a good job on the science was fundamental to making their missions successful. So even if all Apollo astronauts were not exactly in love with rocks, they worked hard on their roles as lunar field geologists. Still, many moonwalkers were out-and-out curiosity seekers, excited by the opportunity to pull back the curtains on the Moon's mysteries. Can we reignite that spirit of curiosity and adventure?

EARTH'S MOON IS EVERLASTINGLY sprinkled by footprints and other odds and ends from the Apollo lunar landings. Why should we return? The answers have attracted global attention, carved out by experts from the world's space agencies, taking on the form of a Global Exploration Strategy.

In broad terms, extending a human presence to the Moon enables eventual settlement. From a perspective of sheer scientific knowledge, scientists could address fundamental questions about the history of Earth, the solar system, and the universe—and about our place in them. Returning to the Moon means testing technologies, systems, flight operations, and exploration techniques to reduce the risks and increase the productivity of future missions to Mars and beyond. A vibrant space exploration program engages the public, motivates students, and helps develop the high-tech workforce that will be required to address the future challenges facing humankind.

Reconnecting humans with the Moon provides a challenging, shared, and peaceful activity that unites nations in pursuit of common objectives. Doing so expands Earth's economic sphere with benefits to life on the home planet.

Perhaps the first man on the Moon said it best, testifying on May 26, 2010, before the Science and Technology Committee of the U.S. House of Representatives. "Some question why Americans should return to the Moon," Neil Armstrong told them. " 'After all,' they say, 'we have already been there.' I find that mystifying. It would be as if 16th-century monarchs proclaimed that 'We need not go to the New World, we have already been there.' " He continued by adding, "Or as if President Thomas Jefferson announced in 1808 that Americans 'need not go west of the Mississippi, the Lewis and Clark expedition has already been there.' "A growing cadre of space strategists, engineers, scientists, architects, and others are following in the footsteps of Apollo today, orchestrating a return to the Moon. But the call this time is not just to visit, but to stay.

CHAPTER FIVE

LAYING A NEW FOUNDATION

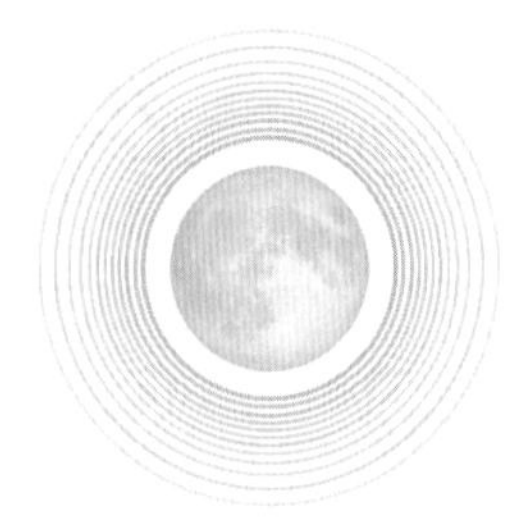

WE LIVE IN A POST-APOLLO world today. The dust the moon-walkers kicked up has long since settled down. The Apollo program and precursor robotic probes spurred a thorough, but preliminary, investigation of the Moon. More recently, orbiters and on-the-spot looks by landers have revealed the lunar countryside to be rich in abundant resources. Some space aficionados optimistically suggest that quadrillions of dollars worth of valuable minerals may exist on the Moon in the form of water, helium-3, and rare-earth minerals. The Moon, in a sense, is an oversize Earth-orbiting space station, a natural satellite outside of Earth's gravity well, rife with raw materials that can be put to practical use.

Speckling its hostile surface and flying above the world are robotic spacecraft, readying the welcome mats for the next round of human adventurers. Some envision putting up arrays of science gear to probe the surrounding universe while others want to use the barren surface as backdrop for collecting solar energy to empower our own planet. And some caution about mauling the Moon, trashing its bleak beauty through aggravated assault: strip mining and off-Earth industrialization. Whatever the case, it is likely that soon the surface of our partner world will see changes from scientists, entrepreneurs, factory owners, and even tourists. But first, we need further rounds of robotic exploration to know what the next series of crewed missions can expect.

Experts say the Moon can be morphed from a desolate orb to a bounty of economically useful resources. Sequestered in polar craters and hidden from the Sun's warming rays, deposits of water ice lie in waiting. The reserves in craters haven't seen sunlight for more than a billion years, locked in cold storage. A series of processing steps can alter that ice into oxygen and water supplies and even rocket fuel. Likewise, the Moon is a storehouse of helium-3, an isotope of helium, that some believe can be harvested into a clean and efficient form of energy. Harboring no appreciable atmosphere, the Moon could also act as a focal point for gathering and beaming solar energy to Earth and other destinations in space.

There are missing puzzle pieces before we can move forward, however. Consider the basics: What's where, in what forms and concentrations? To what degree are the materials that we can utilize intermixed with other materials? How difficult will it be to extract, process, and use volatiles in a situation one-sixth the gravity of Earth? How will mining and processing schemes *actually* perform on the Moon? Laboratory testing of concepts here on Earth is one thing, but the lunar surface poses many challenges, and there are bound to be surprises and snafus.

—||—

DATA SUGGEST THE MOON has water locked away in its icy soil, especially at the lunar poles. The challenge is finding an effective and low-priced way to get it, says Philip Metzger, a planetary scientist at the University of Central Florida's Florida Space Institute. "We don't know if water ice is primarily 'dirty snow,' or if it is gravel-sized chunks of pure ice mixed into otherwise dry regolith, or something else," Metzger says. "We suspect the soil is looser or fluffier at the poles of the Moon than it was at the Apollo sites." It may be difficult to drive on the more loose soil without getting stuck. Right now, he says, there is no relevant experience to design a vehicle capable of working in loose soil and low gravity. This is an engineering and mission risk that can only be addressed by sending rovers to the Moon's polar regions. Extraction methods must be developed. If the soil is very porous, it will be difficult to use thermal extraction methods because heat transfer will be extraordinarily low. On the other hand, permeable soil will make it easier for vapors to make their way out of the heated soil.

The results of these investigations will challenge the idea of strip mining, Metzger observes. It may be much easier to scoop bucket loads of frozen soil, but driving with the mined soil will be difficult unless wheels can be designed to operate in that lunar environment. This all requires ground truthing the process via rovers at the lunar poles. In the past, the idea was to use big equipment to dig for water and then haul ice chunks to lunar processing plants. But now researchers are working on a process that may require less-massive equipment and therefore be more reliable as a way to pull out the water. If we can extract the water in place from the ground, there would be no need to move tons of soil around. That means lighter gear, and weight matters when talking about getting things into space and onto the Moon.

Rocket propellant derived from lunar sources may well be one of the first economically viable uses of space resources. The constituents of water—hydrogen and oxygen, when separated and liquefied—are the most efficient chemical rocket propellants known. This propellant could be stored on the Moon and/or sent to space-based tankers in cislunar space—the space between the Earth and the orbit of the Moon—for potential sale to companies and other countries.

—||—

ROBOTIC SPACECRAFT paved the way for Apollo, and many more robotic craft have returned to the Moon since then to uncover the countless mysteries that our neighboring world holds tight. Each of these trailblazing spacecraft, first from the former Soviet Union and the United States and now from Japan, India, Europe, and China, has been instrumental in unveiling the true nature of Earth's Moon.

The final pair of Apollo astronauts left the surface in 1972, and the last up-close access Earthlings had to the Moon came by way of samples retrieved by the Soviet Union's Luna 24 in 1976. Luna 24 did not carry cameras. It would be 14 years before another spacecraft visited our Moon. The 1980s could be considered the "quiet Moon decade" when no lunar missions by any nation were launched. NASA's primary dollar focus was dominated by readying the piloted space shuttle for its first flight in 1981.

Japan's first lunar mission, Hiten, launched on January 24, 1990, from the Kagoshima Space Center, rocketed into a highly elliptical Earth orbit. The spacecraft whisked by the Moon 10 times that year, testing and verifying technologies for the country's future lunar and planetary missions. By clever use of the spacecraft's thrusters, Hiten would be finessed first into a temporary lunar orbit and then placed into a permanent lunar orbit in February 1993, where it remained until it was intentionally crashed into the lunar surface a couple of months later. Hiten's objective was to measure cosmic dust between Earth and the Moon, a collaborative effort with Germany. Hiten also carried a "grandchild satellite" named Hagoromo, released on March 18, 1990, in the vicinity of the Moon. This small 26-pound satellite was designed to record temperatures and electrical fields around the Moon and radio the data to the mother ship for relay to Earth. However, communication with Hagoromo was lost shortly after its release. As Japan's first ever lunar flyby, lunar orbiter, and lunar surface impact mission, Hiten put Japan on the space exploration map as the third nation to achieve each of these goals.

In January 1994, as America's Clementine mooncraft departed from Vandenberg Air Force Base, the United States got back in the game of lunar exploration. Clementine was a military initiative, run by the Ballistic Missile Defense Organization under the United States Department of Defense. The spacecraft was built and operated by the Naval Research Laboratory (U.S. Navy), and the sensors were designed and built by Lawrence Livermore National Laboratory (U.S. Department of Energy). The objective of the mission was to test sensors and spacecraft components under extended exposure to the space environment. Furthermore, the probe was implementing a "faster, cheaper, better" management approach, moving from conceptual design to launch in only 22 months—a time frame that, for its time, was pioneering.

During two months of circling the Moon, Clementine captured 1.8 million images of the lunar surface. The Clementine imaging experiment showed that permanently shadowed areas do exist in the bottom of deep craters near the Moon's north and south pole. But looking for deep dark secrets within a permanently shadowed area presents a problem: No pictures can be obtained. Instead, Clementine used an investigation technique known as the bistatic radar experiment, using a radio signal transmitted from the craft's high-gain antenna pointed toward a lunar surface target. The signals reflected off the Moon and were received back on Earth for analysis. Frozen volatiles such as water ice are much more reflective to radio waves than lunar rocks are. NASA's lunar prospector later confirmed the Clementine findings.

When Clementine departed the Moon in May 1994 for a flyby encounter with the near-Earth asteroid, Geographos, an onboard computer malfunction caused a thruster to fire until it had used up all available fuel. That left the spacecraft spinning at about 80 rpm with no spin control and ended the mission.

On the water trail tale following Clementine was NASA's Lunar Prospector, launched January 6, 1998, from Cape Canaveral and into low polar orbit, from which the craft mapped the lunar surface using only scientific instruments, no cameras. From January 15, the drum-shaped satellite spent a year charting the entire landscape of the Moon from an altitude of about 60 miles. Early data returns from a neutron spectrometer indicated significant amounts of water ice at the lunar poles. Neutron detection is extremely sensitive to the presence of the hydrogen atom, and a significant amount of hydrogen would indicate the existence of water. Resulting data suggested that as much as an estimated six billion metric tons of water ice were mixed into the lunar regolith at the north and south lunar poles, indirect evidence in accord with Clementine's findings for the south pole.

In extended mission mode, beginning in January 1999, Lunar Prospector's orbit was reduced to 18 miles altitude, then later down to a scant five miles above the lunar surface. With instruments brought closer to the Moon, chiefly those built to tease out more information on weak localized magnetic fields, spacecraft sensor data were greatly enhanced.

Thanks to the workhorse operations of Lunar Prospector, scientists assembled the first global maps of the Moon's elemental composition. They also performed key measurements of lunar outgassing events: the release of gases that percolate up from deep inside the Moon. Seven new mass concentrations of high-density material in or below the lunar crust were revealed, three on the near side and four on the far side.

In late July 1999, the Lunar Prospector mission ended, the craft's trajectory deliberately aimed for a permanently shadowed area of a crater near the lunar south pole. Scientists had hoped that a collision there would release water vapor from the suspected ice deposits. Earth- and space-based observatories including the Hubble Space Telescope were called upon to look, but no plume was observed.

THE FIRST YEARS OF THE 21ST CENTURY saw a new wave of lunar exploration, with countries and agencies never before involved now sending robotic craft to the Moon. In the early 2000s, the European Space Agency—an international organization with 22 member states—launched its first spacecraft moonward. Known as SMART-1 (for Small Missions for Advanced Research in Technology), it took a leisurely pace to travel to the Moon, testing a new xenon-fueled ion engine. Repeated burns of the ion engine gradually spiraled the craft outward from Earth to the target, and as it approached its destination, in a physical process called lunar resonance, the craft used the tug of the Moon's gravity to widen its spiral orbit.

In mid-November 2004, after 14 months of this slow but fuel-saving space travel technique, SMART-1 slipped into lunar orbit. Over the following 20 months, SMART-1 performed a number of scientific tasks. SMART-1's camera allowed scientists to investigate the Moon's topography and surface texture. Infrared and x-ray spectrometers scoured the lunar surface, using over 250 wavelength bands to map the surface distribution of minerals, such as pyroxenes, olivines, and feldspars.

SMART-1's ion engine was shut down in September 2005 after it had exhausted its propellant supply, and the mission ended a year later with the orbiter performing a controlled crash into the Moon.

In early October 2007, the Japan Aerospace Exploration Agency (JAXA) made its way into lunar orbit with Kaguya, the largest lunar mission since the NASA Apollo program of the 1960s and '70s. This multifaceted project consisted of three satellites including an orbiter with scientific instruments, a very long baseline interferometry radio satellite, and a relay spacecraft tasked to transmit signals when the orbiter was on the far side of the Moon, out of direct contact with the Earth: a strategy to help analysts estimate the Moon's far-side gravitational field.

The Kaguya main orbiter performed 10 months of lunar science operations, and JAXA teamed with NHK (Japan Broadcasting Corporation) to gather the world's first high-definition imagery of the lunar surface. The mission obtained data on the Moon's mineralogical composition; its topography, geology, and gravity; and the lunar and solar-terrestrial plasma environments. Using a radar sounder, the Kaguya mission confirmed the existence of multiple large lava tubes underneath volcanic regions of the Moon—underground voids that could possibly offer secure subsurface shelter for humans and scientific instruments.

Kaguya concluded its two-year mission in the same way many other lunar missions have, by being sent careening into the lunar landscape. The crash site was in darkness at the time of the impact, but infrared wide-field camera and spectrograph equipment back on Earth detected the flash of the high-speed plunge of hardware.

India's first robotic Moon mission, Chandrayaan-1, rocketed onto the lunar scene in November 2008, when it entered into lunar orbit. Developed by the Indian Space Research Organization (ISRO), the spacecraft put to the test India's technological capabilities as it sought to return scientific information about the geological, mineralogical, and topographical characteristics of the Moon. Some of the instruments it carried were provided by the United States and the European Space Agency—an example of how global collaboration, with multiple nations contributing experiments and sharing data, has increasingly become the norm in recent years.

Chandrayaan-1's planned two-year mission was cut short, however, when contact with the craft was lost in August 2009.

Prior to that loss of signal, the orbiter had successfully hurled an India-built 75-pound Moon Impact Probe (MIP) equipped with a video camera, a radar altimeter, and a mass spectrometer into Shackleton crater at the Moon's south pole. A Moon Mineralogy Mapper provided by NASA helped detect water molecules on the lunar surface. As planned, the MIP underwent a 25-minute controlled fall, during which its onboard spectrometer detected an above-surface cloud of water molecules. After that, it struck the lunar terrain at 3,800 miles an hour—an intentional end to the probe.

Amazingly enough, Chandrayaan-1 came back into view in March 2017, when an interplanetary radar technique developed by scientists at NASA's Jet Propulsion Laboratory detected the derelict spacecraft still running rings around the Moon.

WITHOUT QUESTION, NASA's long-lived Lunar Reconnaissance Orbiter (LRO) has been a game changer in the ability to inspect the Moon. On the job since entering orbit around the Moon on June 23, 2009, LRO is still making fundamental scientific discoveries. The next humans to visit the Moon will know precisely where to land courtesy of LRO imagery.

There's likely no person on Earth who has accumulated more day-to-day face time with the Moon than Mark Robinson, principal investigator at Arizona State University of the LRO Camera (LROC). Over the years, LROC systems have yielded vivid imagery from the Moon—imagery that has only intensified Robinson's reactions. "If anything, it's more beautiful and more mysterious . . . and even more inviting than it has ever been," he says. It may look cold, beat-up, lifeless, and foreboding to some, but the LROC images prove otherwise.

Two narrow-angle cameras on the LROC produce high-resolution, black-and-white images of the surface, capturing photos of the poles with resolutions down to about 3.3 feet. A third wide-angle camera takes color and ultraviolet images over the complete lunar surface at an almost 330-foot resolution. Images from all the cameras identify potential resources for human crews and safe landing sites for both robotic and crew-carrying spacecraft.

LROC has been used for mapping both the permanent shadow and the sunlit regions on the Moon. It has also helped estimate the impact rates of objects smacking into the lunar surface over the past decade. The Moon's exterior is predominantly shaped by impacts of cometary and asteroidal materials—a process that continuous. Since LRO has operated in Moon orbit for so long, the craft loops back over the same region time and time again, snapping images of the same area. These surface pictures show the rate of new lunar crater production and provide a striking picture, so to speak, showing how pelting micrometeorites create craters and modify the lunar surface.

By combining LROC imagery and historical data, cartographers have been able to create detailed maps of the Apollo landing sites. High-resolution imagery of the six landing spots, from Apollo 11's 1969 landing to Apollo 17 in 1972, reveals the lunar module descent stages sitting on the Moon's surface, left behind by the departing astronauts, as well as lunar surface experiment packages and parked rovers. Faint trails of the astronauts' footprints show up, including observable tracks from the last three Apollo landing excursions as those crews rolled across the Moon's surface in their rovers.

The LROC images even show American flags still standing and casting shadows at all the sites, except for Tranquility Base, the Apollo 11 landing site. The rocket blast as Armstrong and Aldrin departed the scene may have knocked over that flag. Some speculate the flags have been bleached white by the Sun's harsh ultraviolet radiation, but no one really knows the color or condition of the flags at this point. Future expeditions to those sites will find out.

—||—

THE MOON'S GEOGRAPHIC FEATURES are not the only characteristics we need to know about to return to the lunar surface. The magnetic environment is also crucial, in particular how the Moon responds to the radiation flares from the Sun, which can have devastating effects on not just equipment but also human health.

In 2011, two repurposed satellites arrived in lunar orbit, maneuvered there from Earth orbit. Originally, they were members of a NASA five-probe THEMIS (Time History of Events and Macroscale Interactions during Substorms) mission, which began as a two-year-long project using spacecraft to study the physical processes in near Earth space that trigger auroras in our planet's magnetosphere. Sent to the Moon, the project earned a new name—Acceleration, Reconnection, Turbulence and Electrodynamics of the Moon's Interaction with the Sun (ARTEMIS)—a fitting acronym, since Artemis was an ancient Greek goddess, a virgin huntress associated with the Moon. The two spacecraft were deployed to study how the solar wind, with its high-energy radiation, interacts with the Moon. The radiation that strikes the Moon varies from large fluxes of solar wind ions to galactic cosmic ray particles. Without a strong magnetic field of its own or a thick atmosphere, these energetic particle events from the Sun pose serious radiation hazards for humans and equipment.

In 2011 NASA delivered another set of twins to the Moon: the Gravity Recovery And Interior Laboratory (GRAIL) spacecraft. Named Ebb and Flow, this twosome was designed to precisely measure and map variations in the Moon's gravitational field, sensing gravitational tugs as they orbited the Moon—pulls coming from both visible features, such as mountains and craters, and from masses hidden beneath the lunar surface. This same technique revealed differences in density of the Moon's crust and mantle and helped answer fundamental questions about the Moon's internal structure.

GRAIL data shed new light on the formation of a huge bull's-eye-shaped impact feature on the Moon: the Orientale basin on the southwestern edge of the Moon's near side, barely visible from Earth. Formed about 3.8 billion years ago, the basin's most prominent features are three concentric rings of rock, the outermost of which has a diameter of nearly 580 miles. GRAIL data helped to resolve the size and location of the initial depression created when impact first dislodged surface material to form the crater, and determined that it measured between 200 and 300 miles across. Any recognizable surface remnants of that crater were erased by the aftermath of the collision.

The GRAIL mission generated the highest-resolution gravity field map of any celestial body. This map revealed an abundance of features never before seen in detail: tectonic structures, volcanic landforms, basin rings, crater central peaks, and numerous simple bowl-shaped craters. GRAIL confirmed that lunar "mascons"— enormous concentrations of mass that hide just below the surface and tug on spacecraft in unexpected ways—were generated when large asteroids or comets impacted the ancient Moon, when its interior was much hotter than it is today. The results certainly confirm that the Moon's gravity field is unlike that of any terrestrial world in our solar system.

Because of GRAIL's ability to map the Moon's uneven gravity field, future robotic and human spacecraft can navigate with greater precision to make pinpoint landings. GRAIL data also pointed to vacant lava tubes under the battered surface of the Moon, contributing to the thought that these might provide future underground dwellings for human expeditionary crews.

The GRAIL probes operated in a nearly circular orbit near the poles of the Moon until their mission ended in December 2012. At the conclusion of an extended mission, Ebb and Flow were purposely targeted into crash sites near the Moon's north pole, selected to avoid accidentally plowing into any of the historic Apollo landing sites. Even to the last drop of propellant, the twosome coughed up engineering data. The probes fired their main engines until their propellant was gone, giving spacecraft designers information to accurately forecast fuel requirements on follow-on missions.

The present decade saw several other lunar spacecraft from NASA, such as the Lunar Atmosphere and Dust Environment

Explorer (LADEE). It circuited the Moon for seven months between 2013 and 2014 to collect detailed information about the rarefied lunar atmosphere, conditions near the surface, and environmental influences on lunar dust.

LADEE determined the global density, composition, and time variability of the fragile lunar atmosphere. The Moon's atmosphere is thought to be only one hundred-thousandth the density of Earth's and is considered a vacuum. The thinness of the atmosphere at the lunar surface is comparable to the outer fringes of Earth's atmosphere; there are only about 100,000 to 10 million molecules a cubic centimeter. So thin, in fact, atoms and molecules almost never collide. This miniscule lunar atmosphere extends all the way down to the Moon's surface and is technically called a "surface boundary exosphere."

LADEE confirmed that the Moon's exosphere is made up of primarily helium, argon, and neon, mostly coming from the solar wind. All these elements impact the Moon, but only helium, neon, and argon are volatile enough to be returned back to space. The rest of the elements will always stick to the Moon's surface. LADEE also detected small amounts of water in the exosphere, conceivably a hint about the water cycle that may fill perpetually dark craters at the poles with water ice over billions of years.

LADEE, for the first time, found that meteoroid strikes increase the abundance of two key elements—sodium and potassium gases—within the thin layer of gas surrounding the Moon.

An intriguing LADEE investigation was how dust particles mobilize and transport electrostatically over the entire lunar surface. Scientists believe that electric charges may build up at the boundary where the "daytime" and "nighttime" sides of the

Moon meet, perhaps lifting dust from the surface high into the Moon's atmosphere. In the 1960s, several NASA Surveyor lunar landers relayed images showing a twilight glow low over the lunar horizon persisting after the Sun had set. A number of Apollo astronauts orbiting the Moon also saw twilight rays before lunar sunrise or lunar sunset. This phenomenon, it has been speculated, may be to blame for the degradation of retro-reflectors placed on the Moon's surface.

The work of LADEE and learning about the lunar atmosphere were deemed essential before sustained human exploration of the Moon substantially alters it. With the atmosphere so thin, rocket exhaust and fumes from spacecraft landers could easily change its composition—a loss for space scientists.

In April 2014, LADEE made a nose-dive on the far side of the Moon, crashing into the eastern rim of Sundman V crater, a final resting place later spotted in NASA Lunar Reconnaissance Orbiter imagery. Spacecraft from many nations now have sent back data to help test and refine established models for lunar origin and evolution. We know much more about surface features—things like crater formation from incoming impactors and how lunar elements are distributed. Those discoveries also include where vitally important materials like water ice might exist, accessible on the Moon's surface. But the past couple decades of discoveries don't stop there.

—||—

THE MOON REMAINS TODAY a willing and waiting target, ready to yield more insight and understanding. Why should we go back to the Moon? The scientific merits are palpable. But there's more. Humankind can learn to live and work successfully on another world, while, at the same time, move to expand Earth's economic sphere to encompass cislunar space and the Moon itself. In doing so, the prospect to strengthen existing and create new global partnerships is omnipresent.

For its part, the Moon continues to cuddle its topside and undersurface secrets. And it does so in spite of those seeing it as a case closed, been-there, done-that world.

Entering the Moon game in the next few years is South Korea's Korea Pathfinder Lunar Orbiter (KPLO). The KPLO spacecraft will carry a total of five instruments to lunar orbit—four from South Korea and one U.S.-built instrument. This would be the Korea Aerospace Research Institute's (KARI) first lunar exploration spacecraft, with plans for several future Moon missions on the table. With the orbiter, South Korea intends to develop its own lunar exploration technologies and to investigate the lunar environment, topography, and resources.

The sole NASA-sponsored instrument on the payload is ShadowCam, for observing permanently shadowed regions, or PSRs, on the Moon. Using a high-resolution camera, telescope, and highly sensitive sensors, ShadowCam can detect seasonal changes as well as measure the terrain inside these enigmatic craters. Researchers are hoping to learn how volatiles, such as water, move toward permanently shadowed regions and how they become trapped there. Developed by Arizona State University and Malin Space Science Systems, ShadowCam will provide terrain information necessary for robotic and human exploration of the Moon's poles. Among its attributes, Shadow-

Cam can find out whether high-purity ice or rocky deposits are present inside PSRs.

India is slated to launch Chandrayaan-2, an advanced version of the nation's earlier Chandrayaan-1 mission to Moon. That first lunar mission was a major boost to India's space program, accentuating the country's research savvy and technological ability to explore the Moon. India's first lunar foray carried 11 scientific instruments built in India, the United States, United Kingdom, Germany, Sweden, and Bulgaria. Although Chandrayaan-1's mission was abruptly cut short when contact was lost with the spacecraft in late August 2009, the Indian Space Research Organization greatly benefited by upgrading and testing its scientific and technological aptitude in deep-space exploration.

Once it launches, Chandrayaan-2 will head for the first ever nail-biting try by India at landing a rover near the lunar south pole. Chandrayaan-2 is configured as a two-module system: an orbiter craft module and a lander craft module that carries the rover. The Chandrayaan-2 orbiter's payloads will gather scientific information on lunar topography, mineralogy, elemental abundance, lunar exosphere, and signatures of hydroxyl and water ice. The lander is named *Vikram* after Vikram Sarabhai, the father of the Indian space program. It is targeted for an ancient plain over 370 miles from the Moon's south pole.

Once down safe and sound, *Vikram*'s activities include monitoring for signs of moonquakes and studying the thermal properties of the lunar surface. It is hoped that the lander can survive one lunar day, or 14 Earth days, before bearing ultracold lunar night temperatures. Rolled off the lander, the six-wheel, 44-pound rover carries two science instruments to look at the composition of the lunar surface. Once again, how well it handles the brutal nighttime temperatures is a finding yet to be determined.

—||—

RUSSIA'S ROSCOSMOS SPACE AGENCY is looking to reactivate its lunar exploration program. The European Space Agency is teaming with Roscosmos on a suite of Luna missions spread over the next few years. These projects will put Russia, after a decades-long hiatus, back in the business of robotically probing the Moon.

First up and out is Luna 25, set to land near the Moon's south pole. The Moon's south pole is more and more a favored location for exploration. A confluence of factors make it so. This locale is viewed as an ideal site for a future outpost with access to craters permanently in shadow that could be repositories of water ice and other minerals—vital resources for exploitation by future explorers. Mountain peaks near the pole are illuminated by the Sun for long periods of time, making them just right for soaking up solar energy to power an outpost. Finally, mountains and basins that do not face Earth make the lunar south pole observatory-friendly. Sheltered from broadcast static from the Earth, zones of radio quietness would allow radio astronomy surveys of the universe.

Two years later the Luna 26 orbiter is planned. Luna 27, a lander, will be larger than its predecessor and will also aim for the south pole. ESA-built guidance and navigation technology will guide it to touchdown. Once on the Moon, this spacecraft will deploy a European-built drill to search for water ice and other chemicals at least three feet below the surface—a drill called PROSPECT, shorthand for Package for Resource Observation and in Situ Prospecting for Exploration, Commercial exploitation and Transportation. Operating at temperatures of minus 150°C, the drill's assignment of penetrating the frozen surface won't be easy. PROSPECT includes a miniature laboratory to analyze the materials it digs up to help reveal the Moon's history and indicate whether future explorers could use local resources to establish a base. These resources can help sustain crews on the Moon. But much remains unknown, and not only about potentially usable resources. PROSPECT is intended to also prepare technologies that may be used to extract lunar resources in the future.

This series of Luna spacecraft could well rejuvenate Russia's interest in establishing a permanent lunar base, a long-standing objective that stretches back to the 1960s. More recently, Russia's Roscosmos has said their lunar probes would first scout out outpost locations, followed by landing humans on the Moon in 2030. Roscosmos is a state corporation that ensures the implementation of the Russian government's space program.

China is also pressing forward on a multifaceted Moon exploration agenda. The country's Chang'e program is built upon three phases: circling around the Moon, landing on the Moon, and returning samples from the Moon before 2020. They appear on track in keeping that schedule. The country has already sent Chang'e 1 and Chang'e 2 to circle and chart the lunar surface in 2007 and 2010, respectively. In December 2013, the Chang'e 3 spacecraft soft-landed on the Moon and unleashed the wheeled Jade Rabbit rover that reconnoitered the lunar surface. In October 2014, the Chang'e 5-T1 probe carried out an eight-day circumlunar voyage that, among checklist items, appraised reentry technology and techniques for hauling back to Earth bits and pieces of the Moon.

China's accomplishments have not gone unnoticed by other nations. Each mission is harnessing new skills to augment the country's growing portfolio of deep-space operations. On China's lunar manifest is the first robotic landing on the Moon's far side. This is a two-part mission—one lander/rover and a relay satellite beyond the Moon to link communications between Earth and the far side lander. The relay satellite, named Queqiao—Bridge of Magpies—was launched in May 2018, with the lander set to journey to the Moon in December.

Queqiao is parked at where a gravitational equilibrium can be maintained—a special place called the Earth-Moon Lagrange point L2, about 280,000 miles from Earth. This is the first ever lunar relay satellite at this position, allowing it line of sight with the far side of the Moon to pass on communication signals back and forth from Earth and the Moon. China's Chang'e-4 far side lander and rover will tote payloads for Germany, the Netherlands, Saudi Arabia, and Sweden. But one of the most unique parts of its payload is testing the ability for plants to grow on the Moon. This mission carries a tin of potato seeds and *Arabidopsis,* a small flowering plant related to cabbage and mustard. Also, silkworm eggs are expected to be on board.

The "lunar mini biosphere" experiment is a cylindrical tin, made from special aluminum alloy materials. At just seven pounds, it contains water, a nutrient solution, and air. A tiny camera and data transmission system allow researchers to keep an eye on the seeds and see if they blossom on the Moon. China's

Chongqing University and 27 other universities designed the experiment to properly control the humidity within the biosphere. A tube directs the natural light on the surface of the Moon into the cylinder to stimulate plant growth. The experiment may help accumulate knowledge for building a lunar base and establishing long-term residence on the Moon. Future colonists will need to eat, and if they can grow food on the lunar surface, that would cut down dramatically on the weight of eatables that need to be launched to the Moon. Chinese space researchers tout the investigation as helping to accumulate knowledge for building a lunar base and long-term residence on the Moon.

Chang'e 4 spacecraft is projected to land in the general region of the southern floor of the Von Kármán crater. That far side spot on the Moon is a promising location for sampling impact melt from the South Pole–Aitken basin, the oldest and largest impact feature on the Moon. Determining the age of this basin ranks among the highest priorities in lunar science, because it would help unravel the timing of the oldest and largest basin-forming events on the Moon. That information is essential to any estimate of the collisional evolution of the early solar system.

A bit farther into the future is an even bolder and highly complex Chinese mission, Chang'e 5. This would be China's first lunar sample return mission, rekindling robotic lunar land, grab, and go campaigns since the Soviet Union's last sample return flight in 1976—more than 40 years ago. The goal is to haul back to Earth nearly five pounds of select lunar samples. New samples returned from selected areas of the Moon are gifts that can yield surprising advances in lunar and solar system science. We already know from past experience—whether from former Soviet robotic sample returns or the "hand-carried" specimens Apollo moonwalkers brought back—many crucial scientific investigations cannot be done remotely. Far better is close contact with samples in Earth labs that are stuffed with new, powerful tools to unmask the Moon's history. Hauled homeward to Earth, lunar samples are inheritances that can fuel lunar and planetary science interests for generations.

The reported landing zone for Chang'e 5 is within the northern Oceanus Procellarum, an area also called the Rümker region. Mons Rümker is a circular volcanic complex that is roughly 43 miles in diameter and some 1,640 feet higher than the surrounding mare surface, those large, dark, basaltic plains on Earth's Moon that are formed by ancient volcanic eruptions. This region is covered by a variety of landforms, such as numbers of mare ridges and domes of volcanic origin. On-location looks by the lander are expected to help piece together the story of lava flows on the Moon.

A demanding mission to say the least, Chang'e 5 is composed of four parts: an orbiter, a returner, an ascender, and a lander. That approach is different from the way the Soviet Union's Luna sample return program operated. The Soviets deposited the sample directly into the Earth return capsule, then lobbed the specimens from the Moon. A transfer of the lunar samples in lunar orbit was not required. Also, the Soviet Union selected landing areas on the Moon so they would have a direct return trajectory to the Earth. In China's case, the Moon lander will grab and stash lunar samples in the ascender. After rocketing off the lunar surface, the ascender will then rendezvous and dock with the orbiter, and transfer the collection into the returner. The orbiter and returner both would head back to Earth, separating from each other en route. Finally, the returner craft, stuffed with its collectibles, will reenter Earth's atmosphere and parachute to terra firma.

For more than four decades, no samples have been returned from the Moon by either humans or robots. Robotic orbiters, though, have provided significant new remote sensing data that have aided in making detailed maps of the mineralogy of the lunar surface.

This sample return mission is not the final stage of the Chinese Lunar Exploration Program (CLEP). The Chinese space planners see further lunar exploration, including robotic probing of lunar polar regions. Those areas might be valuable sources of hydrogen, oxygen, and other chemical resources to prolong human stay times on the Moon. And although Chang'e 5 has several science objectives, perhaps the real point of the mission is to test technology useful to prepare for human Moon landings: A crew touches down, explores the surface, then rockets off the lunar surface, rendezvous with another craft, and safely returns to planet Earth. In terms of the Apollo program, sound familiar?

CLEP's undertaking is a robust series of missions designed to explore the Moon and prepare for human exploration missions in the coming decades, says Brown University planetary scientist James Head. He worked on the NASA Apollo program: analyzed potential landing sites, studied returned lunar samples and data, and provided training for the Apollo astronauts.

Head has visited Chinese planetary science laboratories at universities and Academy of Science institutes. "It was clear how engaged these groups were in the mapping of candidate landing sites and in preparation of sophisticated laboratory facilities in anticipation of the return of lunar samples by Chinese spacecraft," he says. They were very interested in Head's experience and in the details of training the Apollo astronauts.

Head has no doubt that the CLEP plan will culminate in Chinese crews scientifically exploring the Moon. One has only to look at the CLEP logo to see that the stylized Chinese character for the word "Moon" is the heart of the symbol, surrounding two footprints.

China's willpower to investigate the Moon is part and parcel of a maturing list of nations eager to reach out and purge the secrets from our celestial neighbor. Ostensibly, China's robotic explorers are setting the stage for the country's likely fourth phase of exploration—one that bridges CLEP and their human spaceflight program. That blending of capabilities may be a comeuppance for America, and likely will result in Chinese astronauts shoving foot into the dusty Moon and establishing a base on the Moon by perhaps 2025—a speculated time frame.

Beyond China and its dedicated, methodical approach to Moon exploration, the larger, global picture is telling: Many nations now have lunar territory in their crosshairs, whether for in-country technological advancement, scientific discovery broadcast on the world stage, possible commercial gain, or to join a prestigious club of nations that have the skills to shoot for the Moon . . . and were successfully able to do so. That ensemble of capability is a wave action of progress that codifies and clarifies a bona fide return of humans to once again stamp footprints on the Moon.

CHAPTER SIX

BACK TO THE MOON

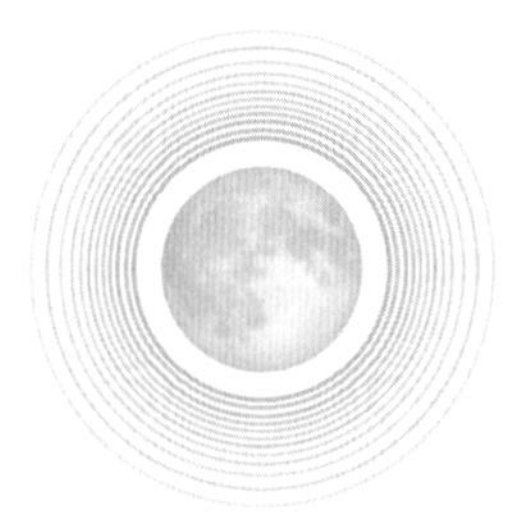

"THE EARTH IS THE CRADLE of humanity, but mankind cannot stay in the cradle forever," commented Russian space visionary Konstantin Tsiolkovsky, one of the fathers of rocketry and cosmonautics. Given that perspective, Apollo was the miracle baby.

Today the Moon is on the right side of the future, a new chapter of our celestial neighborhood watch program and a next phase of solar system science and exploration. "Its proximity to Earth, coupled with reduced gravity and natural resources, make it an ideal location to prepare humans and machines for venturing farther into space," states a 2017 report issued by the International Space Exploration Coordination Group (ISECG), a global community of more than a dozen space agencies united to advance a long-range human space exploration strategy. The new era of human lunar exploration, the document continues, will be able to capitalize on the extensive archive of data sets and imagery collected since the 1970s, all useful in the effort to reduce risk and maximize the rewards that will come from sending humans back to the Moon's surface. "In addition to scientific benefits, human lunar exploration will also contribute to the wider exploration goal to enhance the prospects for humans to live and thrive in space while reaping associated societal benefits."

Six decades of global initiatives have shown that space exploration benefits more than just dipping into the unknown.

Cutting-edge innovation is essential, necessitating advances in science and technology. This requires creating and maintaining a technological workforce, one that is of economic benefit to all spacefaring nations. Hardware and technical competence impact culture, fueling new ways to address global challenges here on Earth.

The ISECG laid out the top-level science objectives that a program of human lunar surface exploration can address. High on that list is a central observation: The Moon preserves the impact record of the inner solar system and yields insights into the impacts on terrestrial planets, specifically on our Earth, that have resulted in upheaval and mass extinction events. Extended lunar exploration could even unlock the mystery of how life on our world originated. The Moon's geologic history provides clues to conditions on early Earth under which life became established and continued through time.

Human operations on the lunar surface can greatly contribute to furthering humanity as a permanently spacefaring species. Many of the resulting discoveries made, technologies refined, and techniques developed will be applied to future efforts to send humans beyond the Earth-Moon system. The science developed for reconnecting with the Moon by way of robots and humans will certainly impact our future. Findings in planetary geology, geochemistry and geophysics, astrophysics, and cosmology, including astrobiology, will not only impact our knowledge base but even shape our shared future as humans consider moving out to other worlds.

Similarly, knowledge in the life sciences will be augmented by pushing the limits of human physiology, psychology, and medicine to safely sustain humans beyond the shores of Earth. The Moon is sure to be an important proving ground to study bone structure changes and bone loss, muscle loss, and changes in the circulatory system due to long-term missions in space under less-than-Earth gravity. Countermeasures to these physical changes will have to be developed, such as exercise, diet, and pharmaceuticals, to preserve crew health.

With one-sixth of Earth's gravity and no meaningful atmosphere, the Moon demands a way of life designed to protect crews from the ruthless onslaught of radiation they will face. We are already aware of the cardiovascular deconditioning, bone density depreciation, and muscle loss the human body experiences when exposed to less gravity than Earth's one-g tug. Experiments and observations over time of crewmembers on the International Space Station have helped plot out the physiological and radiation problems encountered in the microgravity of low Earth orbit, and what measures can be taken to counteract these issues. Long-term residence on the lunar surface will present challenges unique to that environment—a different definition of being "moonstruck."

WITH THE ADVANTAGES of so much robotic research and the important challenges waiting to be faced, why haven't human footprints been imprinted in the lunar dust since 1972? What's the holdup? For one thing, recent American presidential administrations haven't seemed able to stick to long-term goals of crewed exploration beyond low Earth orbit.

Presidential intentions go back decades. Honoring the 20th anniversary of the Apollo 11 moon landing, President George H. W. Bush made a commitment to establish a permanent base on the Moon, to develop capabilities for humans to travel as far as Mars, and to begin what he termed the "permanent settlement of space." The three Apollo 11 astronauts joined President Bush as he made the announcement, and Michael Collins said, "We have rested on our Apollo laurels long enough. It's time to get moving again."

In 2004, President George W. Bush outlined his plan for lunar exploration designed to enable sustained human and robotic exploration of Mars and more distant destinations in the solar system, a series of robotic missions to be initiated no later than 2008. The so-called "Bush Push" also called for the first extended human expedition to the lunar surface as early as 2015, but no later than 2020. Six years later, in 2010, observing that America was no longer racing against an adversary, President Barack Obama announced a space stratagem that called for a set of crewed flights early in the decade of the 2020s to test and prove the systems required for exploration outward from Earth orbit and, by 2025, a new spacecraft designed for long journeys to begin the first ever U.S.-crewed missions beyond the Moon into deep space. "So we'll start," he announced, "by sending astronauts to an asteroid for the first time in history. By the mid-2030s, I believe we can send humans to orbit Mars and return them safely to Earth. And a landing on Mars will follow. And I expect to be around to see it," he added.

Moon advocates were not pleased with President Obama's plan to skip a return to the Moon. The Moon is the most feasible and accessible destination for human exploration beyond orbiting Earth, they argued. Close by and yet a world away, it's a compelling scientific destination in its own right that can dramatically enhance our understanding of the entire inner solar system. Furthermore, Obama's detractors argued, other nations were putting forth significant effort to place their own astronauts on the Moon. Ignoring those endeavors could be detrimental to America.

Obama's shifting of space gears, scuttling an American return to the Moon, was deemed "devastating" by Neil Armstrong. In May 2010 he testified before the Senate Committee on Commerce, Science, and Transportation. "America is respected for the contributions it has made in learning to sail upon this new ocean," he said, and he warned: "If the leadership we have acquired through our investment is allowed simply to fade away, other nations will surely step in where we have faltered. I do not believe that this would be in our best interest."

Fast-forward to the next presidential administration, which yet again changed NASA's crewed space exploration goals. In June 2017, President Donald Trump reconstituted the National Space Council, an advisory group that had existed in years past, created in August 1958 by the National Aeronautics and Space Act that also established NASA. The reinvigorated council was tasked to review current U.S. space policy and long-range goals and to coordinate national space activities, from security to commerce to exploration and beyond.

President Trump's directive underscored the need for reinvigorating America's human space exploration program and called upon NASA to "lead an innovative and sustainable program of exploration with commercial and international partners to enable human expansion across the solar system and to bring back to Earth new knowledge and opportunities." It specifically named a Moon return as a key factor in the program: "Beginning with missions beyond low-Earth orbit, the United States will lead the return of humans to the Moon for long-term exploration and utilization, followed by human missions to Mars and other destinations."

Many experts in the United States agree that long-term bipartisan political backing is essential to progress toward the Moon. "If America is serious about sending humans into space again, a consistent plan must be set with broad, bipartisan support," say researchers with the Center for Strategic and International Studies (CSIS) in Washington, D.C. "Missions to the Moon and beyond will take time and likely extend long beyond the current administration," their statement continues. "The United States needs to set firm goals and stick with them to reduce the risk of cost overruns, to enable commercial and international partnerships, and ultimately to maintain its leadership role in space exploration." Once it is time to plan missions and develop specific technologies to suit either the lunar or Martian environments, progress could be significantly impeded if future administrations continue flipping back and forth between destinations.

Once again, as in 1989 when President George H. W. Bush announced his Space Exploration Initiative, a U.S. president was calling for America to lead a return to the Moon with human explorers. But past initiatives have not been sustained. Will this go-round be any different?

One thing that came out of the 2004 Bush Vision for Space Exploration—and has been built upon since—is the recognition that Earth's Moon is essentially unexplored territory. The six Apollo missions to the lunar surface were much more demonstrations of national prowess than they were well-equipped scientific investigations. There remains an impressive list of things we do not know about the Moon. And the Moon has become an object of interest to a number of spacefaring countries. "It is hard for me to conceive of the United States staying on the sidelines as other countries prepare to send their citizens to the Moon," says John Logsdon, professor emeritus of political science and international affairs at the Space Policy Institute within the Elliott School of International Affairs at George Washington University in Washington, D.C. But the United States needs to make obvious its interest to lead space exploration in the 21st century, says Logsdon. The time for a space race to the Moon has passed; the United States won that race a half century ago. In this respect, our experience with the International Space Station is instructive.

The size of a football field and assembled in segments, the International Space Station has been the most politically complex space project undertaken so far. The first element of the ambitious initiative was lofted in 1998, with the United States, Russia, Europe, Japan, and Canada as lead participants. By mid-2018, over 230 individuals from 18 countries had visited the massive orbital outpost. Continuously occupied since November 2000, ISS is an engineering legacy for the future, a major high point in human spaceflight, and an evolutionary waypoint to deep-space destinations. It is also considered a learning guide for creating a similar multination partnership for exploring and then exploiting the Moon. Over the decades of its development and operation, the nations that have been engaged in building and using the space station have mastered impressive space skills, not the least of which is the ability to work together.

AMID INTERNATIONAL SPACE EFFORTS, the United States is hungry to take the lead, not only through international relationships but by building corporate partnerships as well. NASA is orchestrating an Exploration Campaign meant to galvanize American space leadership by returning to the Moon with commercial and international partners. Shortly after President Trump signed his 2017 policy directive, the space agency began appraising Moon landers offered by the private sector and reviewing its own goals and requirements for lunar explorations. Carrying out the orders of the White House-backed U.S. return to the Moon is James Bridenstine, the 13th and current administrator of NASA. One challenge with Apollo, he says, is that America may have left flags and footprints, but we never went back after 1972. "Ultimately, we want to have landers that go to the Moon with humans on board," he said in 2018. "The goal would be to do that within 10 years. I'm not guaranteeing anything, but that's the objective, that's the goal."-NASA's deep space plans have dual purposes, Bridenstine emphasizes: American leadership in the return to the Moon and in reaching Mars. "We are doing both the Moon and Mars, in tandem, and the missions are supportive of each other." Under this administration's policy, it's not an either-or situation—it's both-and.

Thomas Zurbuchen, associate administrator of the Science Mission Directorate at NASA headquarters, explains that although American innovation will lead the way, partnerships and opportunities with U.S. industry and other nations will be expanded. To his point, service contracts with commercial businesses providing payload delivery to the lunar surface might be signed as early as 2019, and the first of two midsize commercial missions to the Moon, contracted by NASA, could come as early as 2022.

To establish a presence outside low Earth orbit in the region around the Moon, NASA has proposed a Lunar Orbital Platform-Gateway (LOP-G, or simply the Gateway): an orbiting space where people can live, learn, and work around the Moon, and from which they can send missions to the lunar surface. Think of it as a corollary to the International Space Station, but orbiting the Moon.

The Gateway will be incrementally assembled in lunar orbit using the NASA Space Launch System, the Orion crew vehicle, and commercial launch vehicles. Once in place, it will serve as a staging point for missions to the lunar surface and destinations in deep space. Various concepts for its configuration are still on the drawing board, but in general the Gateway will include a power and propulsion element, a habitation module, an airlock to enable science experimentation and crew space walks, a robotic arm, and a logistics segment for cargo delivery.

Crews of four—launched on Orion—would travel to the Gateway on stints initially lasting 30 days and up to 90 days as new modules are added to complete Gateway's full capabilities. Conceptually like the International Space Station, the Gateway is not a techno-replay of ISS, though: It features a minimum of 1,942 cubic feet of livable room, contrasted to the 13,696 cubic feet on the Earth-circling space station. At its fullest, the Gateway will take up 20 percent of the habitable volume of the ISS.

At present, the preferred orbit in which the Gateway would be positioned is a near rectilinear halo orbit, optimized for supporting exploration objectives at the Moon's south pole. That orbit allows the Gateway to make passes close by the Moon and loops farther out while keeping the station within the line of sight of flight controllers on Earth as well as within sunlight, to energize the facility's solar arrays. The Gateway's propulsion system will mainly use high-power electric propulsion, allowing the facility to shift its orbit slightly, depending on what duties are being performed. Some science investigations might need a crew, and others might not. The Gateway will function in autopilot mode as well.

The Gateway's opportunities for science and technology research are numerous, in areas including engineering interactions with the lunar surface through sample return and telerobotics, as well as basic science discoveries in astrophysics, heliophysics, and Earth science.

The U.S. civilian space agency is on a pathway to start construction of the Gateway, perhaps optimistically in 2022, giving the United States strategic presence in the lunar domain. Plans for this spaceport far from Earth are attracting industry and non-U.S. government interest, specifically from Russia and Europe, along with other International Space Station partners such as Japan, which has expressed an interest in working on the Gateway's life support system. The 14 space agencies participating with NASA in the International Space Exploration Coordination Group (ISECG) have reached a consensus regarding the importance of such a facility in expanding human presence to the Moon, Mars, and deeper into the solar system.

"I envision different partners, both international and commercial, contributing to the Gateway and using it in a variety of ways with a system that can move to different orbits to enable a variety of missions," explains William Gerstenmaier, associate administrator for Human Exploration and Operations at NASA Headquarters in Washington. "The Gateway could move to support robotic or partner missions to the surface of the Moon, or to a high lunar orbit to support missions departing from the Gateway to other destinations in the solar system." It could be home for many instruments, he adds, housed within and on the structure itself, or attached to free-flying platforms stationed near the mini-complex.

The orbiting area around the Moon offers a true deep-space environment where we can gain experience for human missions that push farther into the solar system. The not-too-far-away closeness of the Moon translates into days of travel and quick human return in the event of trouble, a significant advantage when compared to other locations, an asteroid or Mars, from which travel must be measured in weeks or months, even years.

NASA is spending significant dollars on the crewed Orion spacecraft and the Space Launch System, the new megabooster rocket that is part of the deep-space initiative. A maiden booster flight—running years behind schedule and caught in costly overruns—appears now slated for 2021, intended to validate the Orion, the launcher, and the ground systems at NASA's Kennedy Space Center in Florida.

At the same time, NASA is counting on the Gateway as a key component in a series of increasingly complex missions. First of the series, Exploration Mission-1 (EM-1) is an unpiloted flight test of Orion that will loop the Moon, flying farther out from Earth than any spacecraft built for humans has ever traversed. It will voyage 280,000 miles from Earth, thousands of miles beyond the Moon over the course of about three weeks. Orion will stay in space longer than any ship for astronauts has done without docking to a space station, then return home, fireballing its way through our planet's atmosphere at 25,000 miles an hour, producing temperatures of approximately 5000°F to test its heat shield, and end with the Orion parachuting to a precision landing off the coast of Baja, California. Success with EM-1 will be a confidence builder, leading to EM-2, a mission with crew onboard the Orion for a complete lunar flyby and return to Earth. Future exploration missions with crew aboard Orion would head for and dock with the Gateway.

In large measure, the Gateway is considered not only as a place to live, learn, and work around the Moon but also as a support system for an array of missions to the lunar surface. Many of the nearly 200 ideas discussed at a recent Gateway science workshop were related to activities on the Moon's surface. For instance, equipment installed on the Gateway could also make whole-Earth observations from space, to duplicate how Earthlike planets orbiting distant stars might appear to future astronomical instruments.

From its unique vantage point, the Gateway can permit astronaut operation of telerobots—semiautonomous robots that can probe the youngest lunar craters from a distance on the

Moon's terrain. Those human-controlled robots can probe the Moon's landscape and do other work. They could stretch out, for instance, radio antennae across the lunar scenery to scrutinize faint radio emissions from the element hydrogen, dating back to when the universe was a tiny fraction of its present age.

"There's a whole new, unexplored world in Earth's backyard," remarks Jack Burns, professor and director of the NASA-funded Network for Exploration and Space Science at the University of Colorado, Boulder. "Lunar far side is dramatically different from regions investigated by Apollo." The far side is the only pristine radio-quiet site to pursue observations of the cosmic dawn, he suggests.

The Gateway can also provide an orbiting retrieval base for human-assisted sample return missions to the lunar far side, a completely unexplored region of the Moon. Studies have already identified high-priority science and exploration sites on the Moon, among them the Schrodinger basin, which is within the South Pole–Aitken basin. In this situation, Gateway crewmembers would telerobotically drive a rover and collaborate with Earth mission controllers as it pursues its scientific explorations. In that type of venture, a robotic ascent vehicle will haul lunar surface samples up from the Moon, delivering them to the Gateway. Astronauts will then transfer the specimens into NASA's Orion spacecraft, which will tote the lunar stash back to Earth for intensive study.

But the Gateway comes at a cost. Solid estimates of dollar numbers to build the spaceport at this point are scarce, but have been ballpark-figured at upwards of $10 billion. Perhaps a shutdown of the International Space Station would provide a wedge of available dollars. Upkeep costs of the ISS are considerable, with NASA spending $3 billion to $4 billion a year on the facility. And the plan for years has been to discontinue the project sometime in the 2020s. Degradation of system components and long-term exposure to the environment of space takes its toll, conceivably making the station an unsafe place to hang your helmet. At some point, decisions will have to be made about what comes next in low Earth orbit.

Ending support for ISS and canceling other NASA projects are seemingly first steps to sniffing out money to build the Gateway and the rest of the lunar architecture. It may be that international and even commercial partners come through to provide some of what's needed. Whether NASA pays contractors directly to build infrastructure or plunks down cash for services purchased through public-private partnerships—either way, the Gateway will still cost major bucks.

Not everybody backs the proposed Gateway. Robert Zubrin, president of the Mars Society, a public advocacy group, considers the Gateway a "boondoggle." With a cost of several tens of billions of dollars, at the least, he says the pricey facility falls short in delivering a useful purpose. His verdict: We do not need a lunar-orbiting station to go to the Moon, or to Mars, or to near-Earth asteroids. We do not need it to go anywhere, in fact, argues Zubrin. "It is true that one could operate rovers on the lunar surface from orbit, but the argument that it is worth the expense of such a station in order to eliminate the two-second time delay involved in controlling them from Earth is absurd. We are on the verge of having self-driving cars on Earth that can handle traffic conditions in New York City and Los Angeles. There's a lot less traffic on the Moon," says Zubrin.

But for many space planners, the Gateway is the way to go. It offers the ability to regain and sustain deep-space exploration beyond the historical but short-duration undertaking accomplished with the Apollo missions.

Considering the important role of the Moon in the future of space exploration and as a focal point for national, international, and private-sector ambitions, advocates say that emplacing infrastructure such as the Gateway in cislunar orbit will extend short- and long-duration human spaceflight opportunities and begin to establish a de facto logistics chain to the lunar vicinity.

Harley Thronson, a senior scientist for advanced concepts in astrophysics with NASA, points out that the idea of a Gateway—or something very much like it—has been bandied about within NASA strategic planning circles for at least two decades. On-orbit upgrade and maintenance of complex science facilities, particularly large space telescopes, is one advantage that has been discussed. Expensive, complex optical systems in free space for astronomy and the Earth sciences could be deployed, serviced, repaired, and upgraded by way of a cislunar space outpost like Gateway, tended by astronauts.

It's true that essential capabilities to keep humans alive during the long voyages to Mars are being developed and demonstrated on the International Space Station. The station is a capacious facility offering plenty of volume, ready access to logistic resupply and repair, and rapid return to Earth in the event of a major problem. Human missions to Mars won't have those luxuries. Consequently, the argument follows, a facility needs to be deployed that will be a more appropriate stepping-stone on the way to the human exploration of Mars. The Gateway could be that while at the same time offering capabilities for achieving important science goals in the vicinity of the Moon, including support for lunar surface operations with both humans and robots.

In a sense, says Thronson, the Gateway is a 21st-century version of the early years of U.S. human spaceflight. After the Mercury single-seat flights from 1961 to 1963, America's Gemini program helped NASA prepare for the Apollo lunar landings as the vital bridge between Mercury and Apollo. The Gateway serves as a bridge between the ISS and the first crewed missions to Mars.

Growing international interest in the Gateway suggests that collaborative cost sharing and bartering for science services may be in the works. A multiagency task force of the European Space Agency, the Canadian Space Agency, and the Japan Aerospace Exploration Agency is studying the engineering requirements for a scenario involving multiple Moon landers and rovers. Another plan details a five-year campaign of five human Moon landings starting in 2028. The Gateway could support any of these undertakings, and scientists are already designing astronaut traverses and rover navigation plans at five different landing sites.

An important advantage of lunar operations from the Gateway has to do with latency—the time it takes for commands to travel from source to destination. A message will take only 0.3 second round-trip between the Gateway and lunar surface.

Compare that with a latency of 2.6 seconds, the best possible round-trip message travel time from Earth through the Gateway and then to the lunar far side. Controlling robots from the Gateway instead Earth, in other words, shaves more than two seconds off each command.

To prepare for surface teleoperation of robots on the Moon, several precursor, trial run experiments have already been done. To reproduce a Gateway-teleoperated lunar rover, three astronauts living on board the International Space Station took turns putting a rover through its paces at NASA Ames Research Center near Silicon Valley in California. Space station crewmembers used interactive tools to control the K10 planetary rover as it moved around in the NASA Ames Roverscape outdoor test bed. These tests represented the first fully interactive, remote operation of a planetary rover by astronauts in space.

Low-latency telerobotics will enhance and extend human reach, enabling astronauts to be present on planetary surfaces without enduring the hardships of actually surviving on those planets. The lessons learned from these lunar operations will feed forward to future low-latency telepresence missions on Mars. It's a logical flow: The road to Mars is paved with local trial-and-error training through cislunar space and on the Moon.

THE U.S.-LED GATEWAY is not the only back-to-the-Moon initiative under way today. There has been mounting interest in Europe to prioritize the Moon as humanity's next deep-space destination. The Moon, European space planners say, can be a springboard to push human exploration out into the solar system, with Mars as the big distant goal on the horizon. European Space Agency officials are embracing the strategic significance of the Moon and pushing forward on lunar exploration missions involving both robots and humans.

Labeling the effort a "comeback to the Moon," ESA planners have been pressing forward on what they have branded a "moon village." Its leading advocate is ESA's director general, Johann-Dietrich Wörner, who sees the village as a product of international collaboration between spacefaring nations and a base for science, business, mining, and even tourism.

Wörner and other space planners propose that the future of space travel needs a new vision. Although the Gateway would provide a travel hub in space, the Moon is the "right place to be," ESA officials say. Still more a proposal than a solid design on the drawing boards, the moon village would be a station, outpost, and base for research, whether by human or robotic means. It is not a village in the sense of a countryside town, with shops and houses, ESA's Wörner explains, but rather a research station with people working and living together, a community of different people with different capabilities, different opportunities. It could draw on an Earth analogue: the Antarctic Station at the South Pole. A Moon station could host a crew of about 10 people, consisting of a blend of staff and field scientists that rotate out three times a year. Autonomous and remotely operated robotic devices support the base.

In principle, it could be a robots-only place, but the idea is to bring together the diversity of spacefaring nations to support astronauts on the Moon along with robots and autonomous rovers at a central location shared by member states and other nations. The moon village could be likened to a mixed-use business park, thriving on the export of lunar products, including information, experiences, and materials. The vision includes industrial-scale operations, a precondition for lunar urbanism that will radically drive technology requirements. But industrial-scale viability cannot be attainable until significant on-the-Moon operations occur. Consequently, governments will need to invest in lunar surface operations.

A moon village is technically viable from a space architecture standpoint, experts say. They point to similar challenges met in building the ISS, which was put together with components from different nations and built by a diverse number of contractors. To be more than a temporary project by multiple governments, though, a moon village must attract private capital. It could provide an opportunity to develop multiple interplanetary businesses, thereby opening a genuinely new economic domain. Far in the future—once humans are traveling beyond the Moon—a moon village might provide refueling pit stop services on the way to other celestial destinations.

A moon village can offer important societal and cultural benefits to humanity, especially if, as visualized, it truly becomes a global undertaking. The Moon is a place where all countries have the possibility to contribute to robotic and human exploration. The moon village is an open concept with the goal of sustaining human and robotic presence on the lunar surface leading to multiple activities, from science investigations (planetary science, life sciences, astronomy) and resource utilization to peaceful cooperation and economic development. In October 2018, another European idea was promoted: a moon race organized by Airbus, a large aerospace firm, along with international partners. The intention is to boost the movement around Moon exploration and enable the demonstration of key technologies required for sustainable lunar exploration.

Managed by The Moon Race NPO gGmbH, a not-for-profit organization based in Germany, the competition invites participants to compete in one of several parallel technology streams: *manufacturing*—to build the first artifact made of lunar resources; *energy*—to survive the Moon night; *resources*—to fill the first bottle of made-on-the-Moon water; and *biology*—to maintain the first lunar greenhouse. The top-performing teams in each category will receive monetary and in-kind rewards.

MOON OUTPOSTS, lunar mining operations, and interplanetary way stations are all visionary and futuristic quests, but an evolving human presence in space means something more than a lunar derivative of Antarctic research stations. Some authorities argue there's need to develop multimission infrastructure that includes utilizing the cislunar region—the space inside the Moon's orbit around Earth—a largely underdeveloped resource. Cislunar space must be developed for its inherent value, not simply because we need to pass through it to get somewhere else. Exploiting this special niche of space will help lessen safety risks when sending humans back to the Moon, to the near-Earth asteroids, and onward to Mars and beyond. A cislunar propellant infrastructure that exploits lunar water and stockpiles and delivers propellant to users at rendezvous points in cislunar space or low Earth orbit could become an important business.

Although public and private stakeholders in spacefaring countries have proposed visions for cislunar development, suggests James Vedda of the Aerospace Corporation's Center for Space Policy and Strategy, no widespread consensus on what to build and how to build it has yet emerged. In the end, putting dollars into cislunar development makes sense as a strategy for boosting space commerce and expanding the human footprint in the solar system. But pulling together a larger, effective space infrastructure will involve a combination of government agencies and private-sector entrepreneurs from around the world. Teamwork between the public and private sectors and across national lines will be compulsory.

The metrics for success should not focus on how quickly we get to Mars or how many people we have living in space; rather, we should be measuring how much we're gaining in capabilities and knowledge, leading to increased prosperity, global solutions, and discovery. If the Moon is treated as merely a springboard—to Mars or to another destination—that's a failure from a policy and planning perspective. But if the Moon is seen not only as a stepping-stone but also a valuable resource for materials and knowledge in and of itself, we're on the path to success.

CHAPTER 7

GIVING THE MOON THE BUSINESS

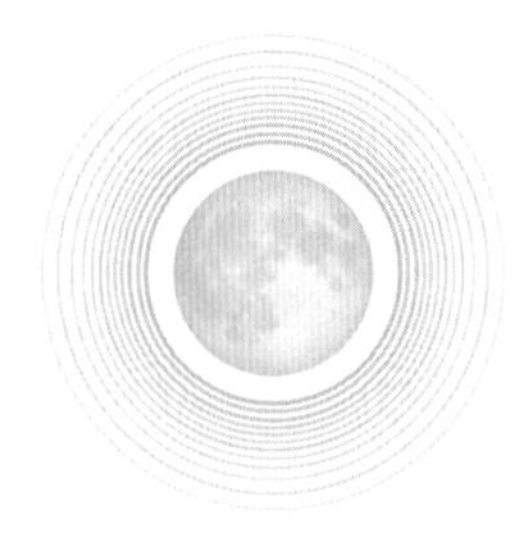

THE HISTORIC EVENTS of the Apollo program, with 12 astronauts walking on the Moon's surface, were exciting, but also difficult and expensive. Today, getting back to the lunar surface remains just that: hard to do and costly. Adjusted for 45 years of inflation, any new program that mirrors Project Apollo will cost several times more. NASA paid for that first program to put humans on the Moon. The U.S. government saw it as a priority, a rite of passage that trumped the former Soviet Union's forceful space intentions. As a result, the space agency's annual nondefense discretionary budget rocketed from $500 million in 1960 to a high point of $5.2 billion in 1965, at which point NASA funding represented 5.3 percent of the federal budget.

NASA's 2018 budget is $19.1 billion, roughly equivalent to NASA's 1963 budget of $2.55 billion, adjusted for inflation. However, as a percentage of federal spending, the space agency's budget is lower than in 1963, let alone 1965, and is actually more like what it was in 1960—about 0.5 percent of the U.S. budget. If we want to put humans back on the Moon Apollo style, that budget won't make it. Public-private partnerships are going to be essential for a return to a robust Moon strategy. This time, a successful Moon agenda will have to include several profitable commercial ventures able to deliver lunar services. "The first lunar missions were designed for space *exploration*. The new ones must be designed for space *exploitation*," says Marshall Kaplan, chief technology officer of Launchspace Technologies Corporation, a respected consulting group working with commercial, government, and military clients.

A growing ensemble of private space groups is ready to take on the task, all prepared to give the Moon the business. Earth's Moon is first and foremost a world flush in resources that can and should be put to use. Although scientists still need to pursue basic geologic prospecting to uncover the extent of those resources, if their work shows that useful materials are extensive, the Moon will certainly be brought into Earth's economic sphere of influence. We need to determine if the Moon offers resources that can be considered "reserves"—a supply of a commodity not immediately needed but available if required. In other words, a major driver of the next great space race will be the commercial development of the Moon.

—||—

IN THE NORTHERN HEMISPHERE on Earth, late summer and autumn, you can glance upward and bathe in the light of a full Moon. Soon after sunset, a wealth of bright moonlight floods the early evening. This is prime harvest season, which is why people ascribe "harvest Moon" and "hunter's Moon" to that full Moon. Those terms can also relate to the bounty of resources on hand for hunters and gatherers on the Moon: a heavenly harvest.

The Moon may possibly offer pay dirt with a rewarding mother lode of resources, but what's really there for the taking, and at what cost?

Recent estimates suggest that mega-dollars worth of valuable minerals exist on the Moon. For example, so-called rare earth elements vital to defense and high-technology industries appear to be available there—elements such as uranium, thorium, and other useful materials. Rare earths are a series of chemical elements found in the Earth's crust that are critical to numerous modern technologies, from consumer electronics like cell phones, computer monitors, rechargeable batteries, and light-emitting diodes (LEDs) to a variety of national defense necessities, such as communications gear. Today, 96 percent of these elements come from China, and that country is increasingly putting the pinch on quotas of such elements out of their country.

Lunar scientists have a good idea how these rare earth elements became concentrated on the Moon as its magma ocean evolved and cooled. But they also recognize that early events disrupted and substantially reorganized the process of mineral formation in ways scientists are still trying to decipher. Although lunar data indicate the presence of some concentrations of these highly desirable minerals, more work needs to be done to map their character and distribution.

The Moon is also a repository of helium-3, a light, nonradioactive isotope that sells for a million dollars an ounce on the secondary market. One of the most significant scientific contributions of the Apollo missions was confirming the presence of helium-3 on the Moon. Thanks to solar wind, the continuous flow of charged particles from the Sun, the Moon has long been bombarded with large quantities of helium-3. Helium-3 is embedded in the upper layer of lunar regolith by the buffeting of solar wind over billions of years.

Demand for helium-3 far exceeds Earth's supply. On Earth this isotope is rare, but it is abundant in the lunar soil: at least 13 parts per billion by weight. Harvested and converted into a clean and efficient form of energy, helium-3 could mean safer nuclear energy in a fusion reactor because it is not radioactive and would not turn out dangerous waste products. The process of using helium-3 for energy also makes by-products, valuable consumables like oxygen and water.

Upwards of 1,100,000 metric tons of helium-3 are estimated to exist on the Moon's surface. Because the lunar topside has been stirred up by meteorite strikes, helium-3 may even be found down to depths of several feet. Harvesting the element will likely require heating the lunar dust to around 1112°F (600°C). Hauled back from the Moon, the stuff could power still-to-be-built nuclear fusion reactors here on Earth. Chinese space researchers are already developing ways to prospect for helium-3 and rare earth elements on the Moon. Ouyang Ziyuan, a cosmochemist and geochemist and the chief scientist of China's Lunar Exploration Program (CLEP), has said that the Moon is so rich in helium-3, he believes it could "solve humanity's energy demand for around 10,000 years at least."

Besides the rare earth elements, more common commodities may turn out to be valuable on the Moon. Scientists think billions of metric tons of water ice sit at each Moon pole. If these estimates are true, and humans can extract and process that polar ice, then the Moon becomes what the Persian Gulf is to crude oil, contends George Sowers, a space resources authority at the Colorado School of Mines in Golden, Colorado, and a former chief scientist and vice president of advanced programs at United Launch Alliance (ULA).

The Moon's water ice resource can be transformed into liquid water for nourishment, into oxygen to breathe, and into rocket fuel to power ascending spacecraft. A cislunar economy could be supported by mining lunar ice for propellant, but the first requirement is to accurately determine the location, concentration, amount, and distribution of ice reserves in the permanently shadowed regions of the lunar poles.

Scientists are working on how to mine the lunar ice found at the poles. So far, one technology option stands out as the best alternative. It's a passive method, with no complex excavation system and a minimum of mechanical components, a thermal mining technique that would use solar collectors located at the so-called "peaks of eternal light" on the rim of polar craters. These highland regions receive sunlight virtually all of the time. Sets of collectors would concentrate the received sunlight and then direct that light to the bottom of permanently shadowed lunar polar craters, where it would melt the ice. Gimbaled mirrors located around permanently sunlit peaks near the craters could capture up to 99 percent of the sunlight reaching the Moon throughout the year. And using direct solar energy transfer into a supercold crater can provide variable heating, allowing managers to control the production rate for storage in tanks.

Besides helium-3 and water ice, the Moon's surface has proven to contain many other elements, including silicon, aluminum, magnesium, and others often used in manufacturing. Analysis has shown that the lunar crust consists of 43 percent oxygen, 20 percent silicon, 19 percent magnesium, 10 percent iron, 3 percent calcium, 3 percent aluminum, and trace amounts of other elements, including 0.42 percent chromium, 0.18 percent titanium, 0.12 percent manganese, with smaller amounts of uranium, thorium, potassium, hydrogen, and other elements.

The Moon not only provides materials; it offers a unique manufacturing locale. Without much of an atmosphere, the Moon is a natural ultrahigh-vacuum environment of the sort required for many industrial processes, such as generating semiconductors and optical coating materials and fabricating thin-film solar cells.

Any on-Moon production process could depend on an electric power system that is constantly repairable and replaceable, through the simple fabrication of more solar cells. One scientist has developed a method that uses molten oxide electrolysis to extract raw materials from the Moon's regolith and make silicon solar cells. In this design, the fine surface dust prevalent on the Moon is processed to extract silicon and metals needed to fabricate the thin-film solar cells, and the ultrahigh-vacuum environment of the Moon proves ideal for vapor deposition of the thin-film structure.

The Moon may also prove an energy source of the future in other ways. Because the Moon harbors little atmosphere to block or blur sunlight, it could well become a focal point for collecting and beaming solar energy to Earth and other destinations in space. For decades David Criswell, now retired director of University of Houston's Institute for Space Systems Operations, has envisioned solar power stations built on the Moon as a way to provide sustainable and affordable electric power to Earth. He likens it to a heavenly equivalent of a wall plug. The airless Moon receives 600 times more solar energy, he calculates, than is needed to power all of Earth, making the Moon a celestial "sunshine state."

In Criswell's plan, large banks of sunlight-gathering solar cells would feed a lunar solar power system. The primary system base pairs, numbering about 10, would be located near the east and west edges of the Moon as seen from Earth, clustered on both sides of the lunar equator. Energy gathered is exported to receivers on Earth via a microwave beam. These beams are of low intensity, about 20 percent that of natural sunlight, making them safe for humans. Once on Earth, that energy is converted to electricity and fed into the local energy grid. This project can be scaled up to supply the terawatts of electricity required by all 10 billion people estimated to inhabit Earth by 2050, says Criswell, providing adequate power to a prosperous Earth while occupying a small portion of the surface of the Moon.

Meanwhile, the Shimizu Corporation, a Japanese construction firm, has a plan to harness solar energy on a larger scale than almost any previously proposed concept. Their ambitious notion involves building a belt of solar cells around the Moon's 6,800-mile equator, converting the sunlight to microwave/laser energy beamed at Earth, and then converting those beams to electricity at terrestrial power stations. The Luna Ring plan, the company says, could meet the entire world's energy needs.

ONE OF THE LOFTIER GOALS related to Moon exploration includes the space elevator: a thin vertical tether extending from the Moon's surface into space, spanning over 155,000 miles. The plan is to create a permanent tether system: reusable, replaceable, and expandable. The landing site of the tether would be secured at Sinus Medii—Central Bay—a small lunar mare at the lunar equator. The other end would beat the L1 Lagrange point, one of the balance points where the combined gravitational forces of the Earth and the Moon pull equally on an object and keep it in space. The system would be designed to allow for the transport of material along the length of the tether, between the Moon and the environment of cislunar space. A solar-powered elevator car would travel the length of the tether, delivering cargo, thus eliminating the need for entire spacecraft systems (with their landing gear, descent engines, and propellant) to travel to and from the surface of the Moon. It would result in a much easier and less expensive transport system than maneuvering spacecraft to the surface for a soft landing and then a safe takeoff every time.

A station would be located at a parking zone of sorts in the space between Earth and the Moon, where the gravitational pull from both bodies equalizes and would stabilize a space station with little fuel. That operations station would be a busy place, handling cargo shipped to and from the Moon, serving as a refueling depot, and performing zero-gravity manufacturing duties.

The LiftPort Group of Seattle, Washington, says the technology to build this space elevator is already available. Designed with Sputnik-like simplicity, LiftPort's futuristic lunar space elevator has been under study for nearly two decades. Engineers first were working with the concept of an Earth-based space elevator, but as their sights went moonward, they realized that the Moon's low-gravity grip and airless atmosphere were highly advantageous to their designs.

From early on, engineers have known that a space elevator would require some sort of magic material, extraordinarily light but superstrong, flexible, durable, and manipulable. Materials now under study are the carbon nanotube, diamond nanothreads, and graphene—pure carbon, remarkably strong, far stronger than steel, and yet very low in weight. The materials exist to fabricate an ultra-long composite tether, one that uses a core material wrapped by a blend of other superstrong materials, says Michael Laine, who heads the LiftPort group. The organization is appraising a dozen materials, he adds, working to find the Goldilocks mix that is just right for a lunar space elevator.

Once fully functioning, the lunar elevator could soft-land equipment and people onto the Moon's landscape. Advocates of the project see transport of three dozen people to the Moon a year as attainable in the early years of the elevator's operation. The concept allows mining, harvesting, and delivery of lunar assets to Earth, and it encourages habitation of the Moon, even on the scale of Antarctica's McMurdo Station.

A lunar elevator would reduce the cost of lunar landings threefold, says Charles Radley, CEO of Space Initiatives Inc. and a technical adviser to the LiftPort group. What's more, the numbers work, says Radley. A lunar-situated elevator could recoup the cost of collecting material from the Moon and sending it to Earth after 53 payload cycles and reduce the cost of sample return ninefold.

Aside from the technical, managerial, and financial challenges of a lunar space elevator, there is a strategic problem. The physics of the situation calls for a Moon-situated space elevator to sit smack-dab in the middle of the lunar disk, on the equator—and yet the first useful, economically precious deposits of lunar materials are likely to be water ice, to be harvested from shadowed craters near the poles. So how can water ice mining operations at the lunar poles connect to the space elevator's anchor station at the equator? In a final report to NASA on the lunar space elevator notion, researchers proposed two paths. One is to build a tramway from the poles to the equator, like a set of rails here on Earth that forms the route for a streetcar. The problem is, this lunar tramway must be suspended from crater rim to crater rim and will extend six times as far as any Earth counterpart, going the full distance from the pole to the lunar equator, nearly 1,700 miles.

Another approach is to bypass the equator-based space elevator altogether and build lunar slings at the poles to hurl lunar ice, minerals, and regolith to the L1 Lagrange point, where spacecraft would catch and load the materials onto cargo vehicles for the trip to low Earth orbit.

Yes, all speculation. But what's a future for?

Mining materials on the Moon will have to be part of any plan to go back there. Extracting gases and liquids from the Moon will be vital for sustaining off-Earth human settlements, but the technology won't evolve until there is an economically driven demand for oxygen and water, a need that can shore up an investor-base business to employ those resources. One person who recognizes this economic reality is Harrison Schmitt, the lunar module pilot on the sixth and last human Moon landing, in 1972. A professional geologist, Schmitt departed his NASA astronaut duties in August 1975, to run for the U.S. Senate. He served a six-year term as one of New Mexico's senators. Eager to see a rebirth of the zeal for U.S. human space exploration, Schmitt is a staunch supporter of the private, investor-based return to the Moon and the efforts to extract helium-3 for energy production, to use the Moon as a stage for science and manufacturing, and to establish permanent human outposts on the Moon, furthering humankind's sense of destiny as explorers.

Schmitt and others view the Moon as a private-sector opportunity, so NASA can spend its research and development dollars in technology to send humans to Mars. Rocketing humans back to the Moon appears an indispensable next step if we want to increase the probability of success and maximize the scientific return from a future human exploration program to Mars, Schmitt believes. If the private sector can succeed in developing energy via the Moon's helium-3 supply, a great deal of the technological capability to get to Mars would have been developed, he asserts.

The Moon represents a huge economic potential, every bit as promising as any new frontier humans have opened to exploration, inquiry, and discovery, says Dylan Taylor, a leading space investor, philanthropist, and global business executive. As founder of Space for Humanity, he points out that the Moon is the "central outpost" for forays deeper into space. Yes, we have been to the Moon before, but this time commercial groups must put resources into exploration to expand Earth's economic sphere to cislunar space. Science, lunar habitation, and commerce all add up to a space exploration trifecta.

—||—

EVEN IN THE CONTEXT of commercial ventures to the Moon, governments will need to play several important roles in a new lunar economy, asserts space resource expert George Sowers, recently retired from United Launch Alliance. Leaders and negotiators will be tasked with establishing a framework of rights, enforcing contracts, and providing security. Governments will invest in basic science and technologies, as well as spending money in infrastructure through public-private partnerships in large-scale operations such as water mining. Finally, the government should participate in the lunar economy as a customer.

To make his point, before he retired, Sowers became the first person to offer to buy propellant in space. His precedent-making bid set a price for propellant bought either on the lunar surface or in Earth orbit. As a customer, he announced, ULA would be willing to pay about $1,360 a pound for propellant delivered into low Earth orbit. The going rate for fuel on the surface of the Moon was calculated to be $225 a pound. Having a source of propellant in space benefits anybody going anywhere in space, in Sowers's view, especially if you can purchase propellant in orbit for less than it costs to ship it there from Earth. At those offered prices, the cost of any activity beyond low Earth orbit decreases dramatically. For example, this would reduce the cost to deliver mass to the surface of the Moon by a factor of three. It would have a similar effect for a Mars mission.

Sowers testified before the House Committee on Science, Space, and Technology in September 2017, outlining his perceptions of private-sector exploration of the Moon. Space resources are plentiful, he asserted, and utilizing them will free human progress from the resource constraints of Earth. In recorded human history, there have been two major economic revolutions, he recounted: the agricultural revolution of 10,000 years ago, which gave birth to human civilization, and the industrial revolution of 300 years ago, which gave rise to the tremendous increase in human well-being and prosperity we enjoy today. "Space resources will be the third major economic revolution and will usher in an era of unprecedented prosperity and flourishing," Sowers concluded. Space resource utilization can save humankind as it unchains human progress from the constraints of Earth's ever diminishing resources. It was a clarion call championed by the advocates of a return to the Moon.

Some people see the Moon not as an austere companion to Earth but as an orb of profit-making potential. Because of the Moon's proximity, private groups see money to be made. But looming in the wings are voices that suggest a moneymaking Moon might be eclipsed by the lens of law. That reality is prompting legal debate regarding entrepreneurial start-ups setting up shop there. As expected, space lawyers are on hand to give advice on property rights—and wrongs.

At the heart of the matter is the question of whether investors will cough up the funds to build industry on the Moon absent a legal framework. Will private companies be allowed to cash in on the Moon, and if the answer to that question is unclear, will they even venture to do so? Would resource mining be permitted, and what legal protections would exist? To assure freedom of enterprise beyond Earth orbit, U.S. industry needs a regulatory framework that meets our obligations under the UN Outer Space Treaty but still provides a structure with maximum certainty and minimal regulatory burden. Making the Moon a for-profit place of business raises many questions now piquing the interest of outer space lawyers. To what extent can private firms establish an economically viable operation? What laws hamper or promote turning the Moon into a financial engine?

—||—

SOME LUNAR EXPERTS raise cautionary flags about how lunar resource harvesting is unfolding. As we enter this new era, they say, we must figure out the best way that these resources might best improve life on Earth, not just fall back on century-old clichés about one economic system dominating another. The capitalist system works as advertised only when the resources are effectively infinite. Currently, the science community is planning missions without taking into account the business and legal communities that are also moving forward to explore—and lay claim to—lunar resources. Perhaps it's time to tackle this impending collision head-on.

Centuries ago, during the time of continent exploration, whoever planted a flag would claim ownership. Some entrepreneurs are urging similar behavior on "islands in space," such as the Moon or an asteroid. That 16th-century paradigm in the long run led to greater separation of the "have" and "have-not" nations and families, and ultimately led to the international warfare that continues today to control the resource-rich parts of the globe. As we envision our relationship with the Moon, shouldn't we take the time to lay out a more realistic and fair framework, to make life better for all?

For some, existing international law relative to outer space, specifically the Outer Space Treaty of 1967, permits properly licensed and regulated commercial endeavors. Under the Treaty, lunar resources can be extracted and owned, but national sovereignty cannot be asserted over the resource area. Some argue that history clearly demonstrates a system of internationally sanctioned private property, consistent with the Treaty, and encourage commercial lunar settlement and development rather than the establishment of a lunar commons, as envisioned by the largely unratified United Nations Agreement Governing the Activities of States on the Moon and Other Celestial Bodies (or Moon Agreement) of 1979.

It's worth reviewing the history that went on between those two United Nations declarations. Legal experts deliberated the Moon Agreement from 1972 to 1979, and the UN General Assembly adopted it in 1979. But it was not until June 1984 that a fifth country, Austria, ratified the Agreement, allowing it to enter into power in July 1984. As of January 2018, only 18 states have ratified it. Conversely, some of the biggest players in space exploration—the United States, the Russian Federation, and the People's Republic of China—have neither signed, acceded, nor ratified the Moon Agreement.

Space industrialization and colonization advocacy groups of the 1970s, namely the L5 Society that promoted expansive space colonization ideas, successfully led a campaign to oppose ratification of the treaty by the United States Senate. The Moon Agreement "makes about as much sense as fish setting the conditions under which amphibians could colonize the land," said Keith Henson, a key leader in the L5 Society at the time. If the United States backed the Moon Agreement, members of the society insisted, its provisions on the exploitation of natural resources would discourage private investment in space, stall the mining of extraterrestrial materials, slow the growth of economic activity in space, and, overall, delay the time when large numbers of people could freely live and work in space.

As it now stands, the 1979 Moon Agreement reaffirms and elaborates upon many of the provisions of the earlier Outer Space Treaty, stating specifically that the Moon and other celestial bodies should be used exclusively for peaceful purposes, that their environments should not be disrupted, and that the United Nations should be informed of the location and purpose of any station established on those bodies. But the agreement also provides that the Moon and its natural resources are the "common heritage of mankind" and that an international regime should be established to govern the exploitation of such resources when such exploitation is about to become feasible.

The test of the Moon Agreement, both as treaty and customary law, will not come until the exploitation of extraterritorial resources becomes technically and economically feasible. That's the view of Michael Listner, attorney and founder of the private firm Space Law & Policy Solutions of East Rochester, New Hampshire. When that time comes, though, will the shadow of the Moon Agreement have grown sufficiently to blanket parties and nonparties alike under the penumbra of customary international law?

Any legal setting for the "use" of the Moon is going to be framed in the context of the international law in effect at the time. The one wild card is the Moon Agreement, which although having sufficient ratifications to be international law, has no teeth at the moment, given the parties who have not agreed to it. Whether that changes in the future remains to be seen. "It is one thing to speculate what the legal setting will look like," Listner advises, "but we really won't know what is going to be needed until operations on the Moon begin in earnest. It's easy to theorize what form the legal setting will take but quite another to see what the pragmatic needs will reveal themselves to be."

THE APOLLO PROGRAM spawned technology revolutions in automation, computerization, and communications, and those improvements have led to today's private space organizations. Some call SpaceX, Virgin Galactic, Blue Origin, Bigelow Aerospace, and many others less well known—the "children of Apollo." This new face of space is sustainable as long as these organizations generate profit, can continue to engender enthusiasm, and are willing to continue funding private-sector space. This new money can also fill a void in periods of time when the government gets whipsawed in its budgetary cycles and overtaken by politics and the competing needs of the nation.

If America can put a man on the Moon . . . why can't we put a man on the Moon? That's a question posed by Charles Miller, president of NexGen Space LLC and a former NASA senior adviser on commercial space. Over the last 45 years, NASA made major stabs at re-creating the magic of Apollo. U.S. presidents have in the past tried to lock in on a post-Apollo human return to the Moon: President George H. W. Bush and his Space Exploration Initiative in 1989, followed by President George W. Bush in 2004 with his Vision for Space Exploration, both great plans that went nowhere due to lack of political push and requisite funding. Can the private sector do any better?

Miller served as principal investigator of a study released in 2015, funded by NASA, to assess the economic and technical viability of an "evolvable lunar architecture," a term that embraces the practice of designing and building reliable, comfortable, and sustainable environments in outer space. Doing so is a challenge, one that attempts to maximize the safety of crews in the extremes of space—like living on the Moon.

Miller's report took into account commercial capabilities and services that are existing or likely to emerge in the near term. It stated that with commercial partnerships, America could return humans to the surface of the Moon some five to seven years after being granted the authority to proceed within NASA's existing human spaceflight budget, at a cost of roughly $5 billion for each commercial service provider.

There need to be at least two competitors to keep the price to roughly $5 billion a competitor, Miller explains. Without competition, you get much higher prices, but with competition, the prospects are good. About 10 to 12 years after setting foot on the Moon again, America could develop a permanent base there housing four astronauts. If the commercial industry operates a permanent lunar base, that base could substantially, if not completely, pay for itself by exporting 200 million tons of propellant to lunar orbit for NASA and others to use. The space agency could use it to send humans to Mars, thus enabling the economic development of the Moon at a small marginal cost. A commercial lunar base providing propellant in lunar orbit would reduce the cost to NASA of sending humans to Mars by as much as $10 billion a year. Of course, the space agency needs to devote the up-front resources to build such a base.

Over the past few years since the study was published, the situation has improved, says Miller, pointing to space tech luminaries like Elon Musk with SpaceX and Jeff Bezos with Blue Origin. Both space entrepreneurs are pushing forward with plans to develop boosters capable of reaching the Moon. "We went to the Moon in the 1960s as a race between nations," Miller says. "The best way to go back to the Moon is to set up a race between billionaires."

JEFF BEZOS, the retail billionaire of Amazon.com fame and fortune, is also head of Blue Origin, a company with big plans to pioneer the space frontier. At the age of five, Bezos watched Armstrong, Aldrin, and Collins carry out the Apollo 11 mission. The Apollo program was inspirational for him, helping to fuel his desire and passion to make a difference in space exploration.

"It's time for America to go back to the Moon, this time to stay," says Bezos. Establishing a permanent settlement on one of the poles of the Moon would be ideal, he believes. There, water can be accessed and "peaks of eternal light" in polar regions—mountaintop real estate that's bathed in sunlight more than 90 percent of the time—can provide solar power.

Space travel is extremely expensive, as long as we are using boosters that we toss off en route to space. "We're never going on to do these grand things and to expand into the solar system as long as we throw this hardware away. We need to build reusable rockets, and that is what Blue Origin is dedicated to," says Bezos. "It's a passion, but it's also important." And Bezos is backing up and bankrolling his fervor with Amazon.com profits.

Using seven reusable engines, the Blue Origin's New Glenn rocket would lift off from Launch Complex 36 at Cape Canaveral, Florida. The booster's first stage, equipped with a six-legged landing gear system, is designed to fly back to Earth and land downrange on an ocean-moving platform. Blue Origin has constructed a state-of-the-art facility to build, assemble, and launch New Glenn on Florida's Space Coast. The site will manufacture all new rocket stages, fairings, and adapters; and support fast refurbishment and reuse of each New Glenn first stage. New Glenn will power people and payloads routinely into Earth orbit: a major step toward achieving the Bezos vision of millions of people living and working in space.

Bezos has his sights on the Moon. He is scripting a pathway there via a lunar delivery program called Blue Moon. Likened to his firm's Amazon shipment service, the plan is to send cargo, experimental gear, even habitats to the Moon by the mid-2020s. Blue Moon is designed to be a repeatable transportation service, providing NASA with a commercial lunar cargo delivery solution. A first landing of the cargo craft will be followed by other missions, including snatching samples of lunar ice for delivery back to Earth for in-the-lab study.

Bretton Alexander, director of business development and strategy at Blue Origin, notes that the company's proven technology, developed with private investment and enhanced by NASA's expertise, provides a "fast path" to a domestic U.S. lunar landing capability for medium to large payload delivery. He suggests that one delivery spot could be Shackleton crater, at the lunar south pole—a location that contains ice for fuel and logistics support, mineral compounds for developing structures, and near-continuous sunlight for power generation. Shackleton crater and other locations like it offer pragmatic proving grounds for judging deep-space exploration technologies in close proximity to Earth.

Not to be outdone, Elon Musk, founder and leader of the California-based firm SpaceX, also has company crosshairs on the Moon. "Having some permanent presence on another heavenly body, which would be the kind of Moon base, and then getting people to Mars and beyond—that's the continuance of the dream of Apollo that I think people are really looking for," Musk says.

On September 17, 2018, SpaceX announced that fashion innovator and globally recognized art curator Yusaku Maezawa will be the company's first private passenger to fly around the Moon in 2023. This private lunar passenger flight, featuring a flyby of the Moon as part of a weeklong mission, will help fund development of the firm's once-called Big Falcon Rocket—dubbed a BFR in well-mannered terms—to make the dream of flying to space accessible to everyone. In late 2018, Musk announced the people-carrying vehicle is now called Starship and its rocket booster is simply called Super Heavy.

Explaining his zeal for a Moon trip, Maezawa said he's taking artists with him on the circumlunar voyage. "If Pablo Picasso had been able to see the Moon up-close, what kind of paintings would he have drawn? If John Lennon could have seen the curvature of the Earth, what kind of songs would he have written? If they had gone to space, how would the world have looked today?" He believes that trips to the Moon can be a part of everyone's future. "I will be heading to the Moon," Maezawa says—"just a little earlier than everyone else."

Musk's "Moon Base Alpha" is part of his master plan for Mars colonization. He envisions us as a multiplanetary species in the not-too-distant future, traveling via a SpaceX fleet labeled the Interplanetary Transport System. "We should have a lunar base by now," Musk recently said, speaking during a meeting of the International Astronautical Congress in Adelaide, Australia. "What the hell is going on?" He has reiterated his plans to send people to the Moon and Mars and reinforced his commitment to build and fly a fully reusable two-stage rocket capable of heaving 150 metric tons into low Earth orbit, a booster far larger than the Apollo-era Saturn V launcher.

Musk's space vision came closer to reality with the triumphant maiden flight of his company's Falcon Heavy rocket on February 6, 2018. This potent new breed of reusable monster rocket tossed a 2008-model Tesla Roadster outward from Earth on an elliptical heliocentric route, crossing the orbit of Mars. It was a public relations stunt, yes, but one that showed the potential of the Falcon Heavy rocket, designed from the get-go to reinstate the opportunity to fly crewed missions to the Moon or Mars. After the booming first flight, the entrepreneurial rocketeer reemphasized that his goal was to inspire people to believe, as in the Apollo era, that anything is possible. In that spirit, Musk is moving forward and building the first interplanetary spacecraft destined for Mars.

Based on SpaceX calculations, lunar surface missions can be done with no propellant production on the surface of the Moon. The Musk strategy would have rocket fuel tanks loaded with liquid methane and liquid oxygen to enable regular transport of freight and people to the Moon and return travel back to Earth. Lunar sortie missions could facilitate Moon Base Alpha, an outpost that would in time crank out propellant from lunar resources to fuel outings on to other deep-space destinations.

For Musk, making life multiplanetary is a liberating vision. "You want to wake up in the morning and think the future is going to be great," he says, "and that's what being a spacefaring civilization is all about. It's about believing in the future and thinking that the future will be better than the past. And I can't think of anything more exciting than going out there and being among the stars."

Another billionaire with a different approach for space exploration is Robert Bigelow. His company, Bigelow Aerospace, in North Las Vegas, Nevada, is developing expandable low-cost commercial space habitats for low Earth orbit, the Moon, and beyond.

Ever since Robert Bigelow founded Bigelow Aerospace in 1999, he has dedicated upwards of $200 million of personal wealth to his farsighted space quests. Two Bigelow Aerospace prototype space modules—Genesis 1 and Genesis 2—were orbited by Russian boosters in July 2006 and June 2007, respectively: These expandable modules were the forerunners to the larger human-rated space structures Bigelow is now designing. In 2016, under a NASA contract, the Bigelow Expandable Activity Module (BEAM) was berthed to the International Space Station, further proof-testing the use of expandable volume in space.

Bigelow's Moon initiative centers on launching the company's B330 expandable module into low lunar orbit to serve as a lunar depot. The B330's internal volume is comparable to a third of the current pressurized volume of the entire International Space Station. In October 2017 Bigelow Aerospace and United Launch Alliance announced they are working together on the lunar depot, hoping to build in a public-private partnership with NASA as well. The lunar depot, a strong complement to other plans intended to put people on Mars, could be deployed by 2022 to support the nation's reenergized plans for returning to the Moon. An orbiting commercial lunar depot would provide anchorage for significant business development, Bigelow says, in addition to offering NASA and other governments the Moon as a new location to conduct long-term exploration and astronaut training.

Bigelow Aerospace has drawn up other plans to use habitats to form a base on the lunar surface, declaring that land there has the potential to support economic growth. Throughout the years, Bigelow has advanced a plan for a quick-deploy Moon base capable of housing up to 18 astronauts within inflatable modules, vessels configured to be independent of each other and self-sustaining. The expandable systems are extremely tough, able to rest on any kind of surface. Bigelow claims that such a Moon base would be safe, quick to deploy, and low in cost, with expandable modules able to accommodate an uneven surface. A solar array field could be deployed nearby to power the village of inflatables. "The Moon is a great practice ground," Bigelow suggests, "a very valuable asset for a lot of reasons."

—||—

MOMENTUM IS BUILDING for private enterprise on the Moon. Spearheading private-sector Moon interest, a $30 million Google Lunar XPRIZE competition was rolled out in 2007 to challenge and motivate engineers, entrepreneurs, and innovators from around the globe. The competition signified a new type of space race, far different from the one between the Soviet Union and the United States in the 1960s. The aim of the Lunar XPRIZE was to inspire designers to come up with low-cost methods of robotic space exploration, offering a sizeable reward to push the boundaries of commercial space activity.

To snag the $30 million prize, a privately funded team had to meet three challenges: place a robot on the Moon's surface, explore at least 1,640 feet of lunar terrain, and transmit high-definition video and images back to Earth. Finalist teams could strive for technological milestones along the way, and a slug of extra cash, $4.75 million, was tossed in to help motivate them. These additional milestones included completing one orbit around the Moon, entering a direct descent approach to the lunar surface (to win $1.75 million), or transmitting data proving the spacecraft had soft-landed on the lunar surface.

Early in 2017, XPRIZE officials announced the five finalist teams with verified rocket launch contracts: SpaceIL (Israel), Moon Express (USA), Synergy Moon (international), TeamIndus (India), and HAKUTO (Japan). These teams had until March 31, 2018, to complete their missions. But that vision of a space renaissance came off the rails. By early April 2018, no team had even shot off its lunar lander mission. Following a series of extensions, Google decided not to continue its prize sponsorship, and the XPRIZE halted the competition.

Although the XPRIZE competition had no winner, it did cement sustainable and routine robotic exploration as a goal toward the eventual return and settlement by humans on the Moon. The XPRIZE teams did raise more than $300 million through corporate sponsorships, government contracts, and venture capital, and at least five of the teams came away with launch contracts still in place to land on the Moon's surface in the coming years.

One of those teams is Moon Express, the U.S. entrant to the XPRIZE. "Moon Express is blazing a trail to the Moon to seek and harvest these resources to support a new space renaissance, where economic trade between countries will eventually become trade between worlds," says Bob Richards, founder and CEO of the Silicon Valley–born and -backed enterprise. The group's business plans call for the first commercial lunar sample return by 2020, for scientific as well as commercial purposes. In 2016, Moon Express became the first commercial space company to receive U.S. federal government authorization for a private mission to the Moon. In 2017, Moon Express introduced a family of innovative robotic exploration vehicles designed to collapse the cost of access to the Moon and other deep-space destinations, with regular flights to the Moon planned to begin in 2020.

Astrobotic Technology, Inc.—an early Google XPRIZE contender that dropped out of the competition in 2016—is likewise keen on offering lander services for the Moon, hungry to capture a significant share of a nascent 21st-century space market while helping to get NASA back to the Moon. Astrobotic's product is the Peregrine lunar lander, a craft designed to deliver cargo to the Moon, promising robotic access for companies, governments, and universities for what the group feels is an "industry-defining" price of $1.2 million a kilogram. Company officials say they are on track to begin delivering customer payloads to the lunar surface once a year starting in 2020.

Contracts are already in place, says Astrobotic, to lug small rovers, time capsule artifacts, technology demonstrators, even a Japanese-made sport drink to the Moon. For the first mission, they are including small keepsake items that will be sealed into a single Moon Pod on the Peregrine. After the lunar landing, participants will receive images and videos of the Moon Pod on the lunar surface.

Astrobotic has also partnered with ATLAS Space Operations Inc., a U.S. business specializing in cloud-based satellite management. The deal will develop laser communications dramatically, enabling high-definition video, output from data-intensive experiments, and even virtual reality experiences from the Moon. With a link offering up to one gigabit a second of data to customers, it promises a thousandfold increase in bandwidth for Astrobotic's lunar mission.

THE MOON IS A HOT PROPERTY of the future—and maybe of the past. In fact, one person has plunked down cash for a spacecraft already on the Moon: the Soviet Union's *Lunokhod 2* rover. The honor goes to American entrepreneur Richard Garriott, a well-to-do video game developer. He's a space devotee and the son of scientist-astronaut Owen Garriott who piloted a Skylab 3 mission in 1973 and flew on board the STS-9 space shuttle mission in 1983. Richard himself, in fact, flew a self-financed $35 million trek to the International Space Station in October 2008.

In 1993, during a Sotheby's space auction in New York, Richard Garriott purchased two craft on the lunar surface: the Soviet Union's *Lunokhod 2* rover and the Luna 21 lander, from which the rover had rolled off. That transaction had a hefty price tag for the still-on-the-Moon spacecraft: $68,500. Of course, the rover and lander are not physically in his possession, and who knows how long until they will be. Garriott admits he's in seventh heaven knowing that he's the world's only private owner of an object on a foreign celestial body. In return for his Sotheby's purchase, he received a photograph and a receipt, along with a set of documents in both Russian and English, including a deed of title transfer and certification of ownership.

As the proprietor of *Lunokhod 2*, Garriott believes he also owns that real estate upon which his Russian rover is parked. So far, there's no registered outcry about this assertion. By way of international treaties, nations have agreed not to make territorial claims on the Moon, but the 1967 Outer Space Treaty doesn't even mention private individuals, making this, for some, a gray area. Legal expert Michael Listner responds that it's *not* a gray area, though. The majority view of the probation against sovereign claims within the 1967 Outer Space Treaty is that it extends to private individuals, "which means there is no loophole," Listner says.

Listner says he's familiar with Garriott's "purchase" of the *Lunokhod* rover and his claim that the regolith trapped under its wheels is his personal property. "There is great hyperbole over space resources promulgated by the media and space advocates who do not understand the intricacies of the theory of space resources," Listner explains. "The politics behind it serves only to add confusion to the issue and fuel misconceptions. That said, the argument over the theory of space resources is in its opening phase and is far from settled law."

An interesting Sotheby's side note. In late November 2018, moon rocks headlined a space exploration auction. Described as the only known documented pieces of the Moon available for private ownership, the three tiny lunar soil samples in the mounted display were brought to Earth in 1970 by the Soviet Union's Luna 16 mission. They were originally owned by Nina Ivanovna Koroleva, the widow of Sergei Pavlovich Korolev, director of the Soviet space program, and had previously sold at Sotheby's in 1993 for $442,500. This time they went for $855,000.

—||—

THE MOON IS CHOCK-FULL of vast resources as a result of billions of years of asteroids and comets bombarding it—and, to those of us on Earth, those assets are relatively accessible. Private-sector entrepreneurship is poised to extend Earth's economy to the Moon and generate new opportunities for business markets to support science, exploration, discovery—and merchandising. Of course space experts still need to flesh out if, how, and when the Moon can be integrated into human economic activity. Government-led exploration of space is moving forward, but the question lingers about the private sector's role, and how public and private actors can work together for mutual benefit.

American lawmakers appear to be listening to those wishing to promote business and an economic tie to the Moon. For example, U.S. House of Representatives Space Subcommittee Chairman Brian Babin (R-Texas) sees utilizing lunar resources as a way to ensure U.S. leadership in the future. "Space is so vast and immense that it is foolish to propose that we can meaningfully plumb its depths without the resources, talent, and drive that are so abundant in America's private sector," he has stated.

Will space be the next sector to lead economic growth? Congressman Babin has observed that manufacturing, financial, technical, and philanthropic powerhouses like Andrew Carnegie, John D. Rockefeller, J. P. Morgan, and Thomas Edison dominated the economic and industrial landscape of late 19th-century America. In the late 20th century, Microsoft, Google, Apple, Facebook, and Amazon led advances in information technology. Will we see a prosperous industrial space boom in the 21st century? And if we do, who will be the Rockefellers and Edisons of the future?

CHAPTER EIGHT

NEW MOON RISING

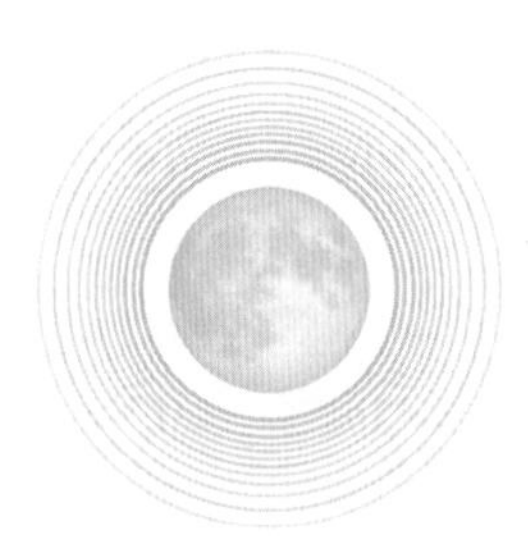

AS WE LOOK INTO THE FUTURE, we see a rekindling of enthusiasm for lunar exploration by robots and humans, and for good reason. The Moon is far from being a "been there, done that" world. Dozens of robots have been to the Moon, we've collected hundreds of pounds of samples, we've landed humans on its surface, and yet the Moon remains largely unexplored territory. Although robotic study can help unravel the mysteries that still await us, it will take human explorers with the proper tools to expose the true nature and origin story of this next-door neighbor in our gravitational grip.

The Apollo program was historically momentous, but it was not a sustainable effort. How we can send the next wave of humans to the Moon in a long-term, stay-put, survive-and-thrive approach remains to be seen. A constancy of purpose is paramount. Minimally, a return to the Moon by humans is liable to mimic scientific outposts at the North and South Poles here on Earth. Arguably, giving legs to the legacy of Apollo means capitalizing on the heritage of that pioneering program. Having no humans follow in the footsteps of the first set of moonwalkers would seemingly make the earlier effort a one-off stunt, devaluing the triumphant achievement.

So what happens now with our planetary partner, the Moon?

The current White House leadership is shaping a new U.S. space strategy, one that reestablishes American interest in the

Moon as first priority. It has also put in motion a policy that will use government and commercial lunar skills and accomplishments to eventually push forward the more distant goal of sending humans to the red sands of Mars.

Several years ago, NASA began collaborative partnerships with private companies. The U.S. space agency is moving forward on the Lunar Orbital Platform-Gateway, which combines American leadership, international space agencies, and private companies. "The Lunar Gateway will create inherent opportunities for commercialization and new discoveries, including unexpected ones," explains Alexander MacDonald, NASA senior economic advisor at the Jet Propulsion Laboratory. By utilizing public-private partnerships as a priority with regard to lunar surface activities, MacDonald reasons, NASA can build on and support the expanding capabilities of intrinsically motivated private-sector individuals and organizations, while also appreciating the international cooperation needed to accomplish all of these projects and missions.

There is increasing global interest in Moon exploration. Dispatching humans back to the lunar surface is not solely an American endeavor. The European Space Agency (ESA) appears poised to work with multiple nations to establish a moon village, and the agency is working with China and Russia to blueprint Moon exploration objectives. China has begun simulations of a bioregenerative life support system for the lunar environment in which animals, plants, and microorganisms coexist. For several years now, in an experiment called Yuegong-1, also known as Lunar Palace 1, male and female volunteers have lived for extended periods of time and worked in a self-contained facility with water and food recycled within the system—an exploratory experience that has long-lasting implications for life on the Moon and planets.

A new space race is starting up in our time—one that, for the most part, China is likely to dominate. This competition is just as meaningful as the lunar rivalry between the former Soviet Union and the United States of the 1960s. The military implications are serious; the economic possibilities real.

The major difference in today's prospect for the return of humans to the Moon, contrasted to Apollo, is our newly developed abilities to use resources on the spot. We still don't know precisely how much, in what quantities, and how accessible those resources are, though. Answers to these questions require substantial work. If we do not develop on-the-spot resources for lunar construction purposes, and if we do not devise efficient methods for making air, water, and rocket propellant out of lunar materials, we short-circuit our ability to live by the Moon mantra of "this time to stay."

In the early days of our lunar return, dwellers on the Moon are sure to be faced with solid work and very little free time. "Think of it like living in a mining boomtown in the old American West," once speculated Paul Spudis of the Lunar and Planetary Institute, for whom this book is dedicated. It will be a scene of solid labor done by hardworking people, with little entertainment, and less-than-desirable food, shelter, and creature comforts. Despite all that, there will be no shortage of people wanting to go there, just as many individuals here on Earth were attracted to the frontier of the American West.

On the immediate "to-do" list: Find water ice, establish a reliable power source, and build up the lunar outpost infrastructure. Grading roads, building structures, and powering up will mean busy hands, busy people. The possibility of 3-D printing structures using the lunar regolith is being assessed around the world. Treated with solar energy, lunar regolith can be sintered and shaped into a variety of usable structures. LIQUIFER Systems Group in Vienna, Austria, for example, has taken part in a European effort called project RegoLight, studying ways to build elements of a lunar base with features necessary for long human habitation, including radiation shielding for inhabited and pressurized modules, shelters that protect machinery from dust and micrometeoroids, and even a Moon pad apron for incoming and outgoing spacecraft.

Give it time, and lunar mining colonies are likely. On the Moon, the most accessible materials will be mined first—those that exist in the lunar topside—to make building materials and other infrastructure items. Water ice in permanently shadowed craters will be mined early on, because it can support human life and be made into rocket propellant. Even Sun-generated electrical power is practical. "These first things mined will gradually be expanded over time to include other commodities, such as metals, ceramics, rare earth elements, and nuclear fuel," Spudis suggested not long ago.

No matter what takes place on the Moon, whether keeping the lights on at a lunar research base or running a mining operation, all functions will require power. Luckily, great strides are being made in building and testing a key energy source for the Moon, a small space-based nuclear power plant. The Kilopower project, funded by NASA, is viewed as a stepping-stone to small fission-powered planetary science missions—including energizing a lunar outpost. The main goal has been to assemble and test an experimental prototype of a space fission power system, a simple approach to the requirements for long-duration, Sun-independent electric power for space or extraterrestrial surfaces. Offering long life and reliability, such a system will produce from one to 10 kilowatts of electrical energy continuously for 10 years or more, explains Lee Mason, the principal technologist for power and energy storage with NASA's Space Technology Mission Directorate.

The proposed power system uses a solid, cast uranium-235 reactor core, about the size of a paper towel roll. Reactor heat is converted to electricity with high-efficiency Stirling engines. One unit would power up a lunar outpost. With this, NASA engineers are striving to give space missions an option beyond radioisotope thermoelectric generators, which provide a couple of hundred watts. A Kilopower unit has the potential to power lander payloads through the lunar night, possibly for months or years. The power level would be suitable to access, extract, and process lunar ice in permanently shadowed craters and demonstrate propellant production. NASA could also co-develop the system with commercial lunar lander companies that supply power to mining ventures or small settlements.

Sending astronauts for long stays on the Moon and to other planets necessitates a new class of power never needed before. A successful lunar campaign using Kilopower technology should be a confidence builder for later Mars missions in which humans will have to depend on the energy-providing system on the planet to power their habitats and manufacture their own return propellant. If successful, a space-rated fission power unit for future lunar explorers becomes a game changer.

—||—

GIVEN THE MOON'S available resources—particularly water, but also potential materials and exceptional places on the lunar surface to set up a base of operations—discussion of the legal framework on divvying up the Moon will assuredly become more heated, requiring resolution as actors from governments and private enterprise establish a presence on the lunar landscape. Any global rush to this newest space frontier brings with it deliberation about property rights, ownership, and exclusivity to extraterrestrial real estate. There's early need to identify and close legal loopholes and forge a set of norms and rules, both nationally and internationally. Laws tend to be built on precedent. Declarations of idealisms—coming in peace for all humankind, ideals of equity and mindfulness of the nonspacefaring have-nots—will all be put to the test.

When eyeing the remote Moon today, perhaps we are underestimating the value of lunar wilderness? If so, might the future of the Moon be a replay of history here on Earth?

Consider this: U.S. Secretary of State William H. Seward signed a treaty in March 1867 with Russia for the purchase of Alaska for $7.2 million. That equates to spending about 2.5 cents an acre for an area twice the size of Texas. (The Moon's surface area is about the size of the continents of Australia and Africa combined.)

Seward's buy of Alaska, envisioned as a way to spread American power throughout the Pacific and encourage American trade and military prowess, was mocked in the press. They labeled the purchase as "Seward's icebox." The acquisition was demonized in the U.S. Congress as "Seward's folly." Congress ratified the treaty by a margin of just one vote on April 9, 1867, and months later, Alaska moved from Russian ownership to the United States.

U.S. settlement of Alaska was a slow-going affair, but the unearthing of gold in 1898 fueled a rapid influx of people to the land. Today, Alaska is a reserve of natural resources, adding to America's affluence and upturning the characterization that the suspect land buy was real estate idiocy.

Is the fascination with our close-by world a parallel?

A leading lunar savant, Wendell Mendell, is a former NASA planetary scientist who retired from the space agency in 2013 after chalking up 50 years of service. His research centered on the geological remote sensing of planetary surfaces, particularly the Moon. Mendell considers our goal of a permanent human presence on the Moon as not a diversion or an impediment but rather part of a historical process. There are many reasons why, he says. A lunar program provides an opportunity to build up space capability in an evolutionary and orderly way, yet at a substantially lower initial development cost than that required to proceed directly to Mars.

Historically, access to a frontier has generated creativity and destroyed old ways of thinking in any generation raised on its threshold. "A well-executed lunar program specifically intended to build experience while returning meaningful science," Wendell believes, "could easily provide the knowledge and confidence—like Babe Ruth—to point to Mars and hit the ball out of the park."

Put succinctly by Clive Neal, a University of Notre Dame planetary scientist and emeritus chairman of the Lunar Exploration Analysis Group: "You can't be a Martian without being a lunatic."

The Moon is a place of learning, as well as a staging area to dive deeper into space.

Reflect on this—a condominium of large, distributed telescopes set up on the Moon. Universe watching in the infrared, for example, would be first-rate, given lunar polar areas that are permanently shadowed. Those localities are some of the coldest places in the solar system. An infrared scope in Shackleton crater at the Moon's south pole offers the best location to build such an observatory. Because infrared telescopes detect heat, that cold location would allow them to operate without background heat to muck up observations.

For a visionary outlook about Moon colonization, ask James Wertz, president of Microcosm, a space mission engineering firm in Torrance, California. In his mind's eye, Wertz sees a colony composed of two or three domes for safety purposes. Each dome is 11 stories tall, topped by 15 feet of lunar regolith. Under each dome, the spacious setting equals two 100,000-seat football stadiums. A low-cost colony would be populated with a thousand people or more roughly 12 years after construction begins.

By that point, science really isn't the goal anymore. Now use of existing lunar resources in a cost-effective fashion is fundamental, like processing Moon material for the structure itself (aluminum and glass) and radiation shielding. Lunarians will even be exporting made-on-the-Moon structural components for space stations, satellites, and space vehicles. At the same time, low-gravity production of pharmaceuticals and semiconductors will be under way. Then there's co-branding, clever marketing of everything from Moon cars to lunar pizza and other out-of-this-world bargains. The lunar colony will also certainly welcome a thriving tourism business.

Wertz concludes that it will take hard work to make all this happen. "Colonizing the Moon is no different than settling California or Alaska, only a bit easier because there is constant communication and resupply," says Wertz. "It's not a 'techie' activity but a people and business activity with a mix of entrepreneurs, service workers, engineers, scientists, explorers, and very busy Maytag repairmen. One thing is certain. If we don't try, we won't get there."

Some see the Moon as a sacred icon that should be protected. Rick Steiner, a conservation biologist, heads the Oasis Earth project in Anchorage, Alaska. Any thought of future strip mining or quarrying the Moon is a troublesome prospect, in his view. Earth's Moon is our closest astronomical companion. It reminds us that we occupy space on a small living planet, one that we all have responsibility to protect and sustain. In a similar vein, the Moon beams upon us, inspires us, and also silently taunts us to take great care in deciding its future.

"The way we think about the Moon is a reflection of the way we think about ourselves on Earth, where our short-term self-interests are literally destroying the biosphere of our home planet," Steiner argues. "Hopefully we can project onto our one and only Moon a more compassionate, sustainable, cooperative future." He and others suggest the Moon should be set aside and guarded from the ravages of industrialization now occurring on our home planet. He points to the imperiled Arctic, one of the most extraordinary, unique, and threatened regions in the global biosphere. Although the Antarctic region is protected by international treaty, the Arctic is not, and it remains largely subject to the commercial and strategic interests of the eight Arctic nations, including the United States. The region is suffering effects of climate change more severely than elsewhere, with growing ecological and cultural consequences. Governments and industry are looking to exploit the Arctic for oil and gas, minerals, shipping, commercial fishing, tourism, and military and strategic interests, all of which may compound environmental impacts and risks.

In 2002, Steiner and his project recommended that the Moon be designated a UNESCO World Heritage site. At the time, this was not procedurally possible as the Moon was not owned by any sovereign state that could propose such designation, but this procedural problem can be corrected. If designated a World Heritage site, the Moon would be the only such site that all people on Earth could see, says Steiner. "Such a designation could be a powerful symbol of unity and cooperation on Earth and that we share a common future on this lovely 'pale blue dot' . . . something we desperately need at this point in our collective history."

MOON RUSH

BUT MIGHT THE NATURALLY foreboding nature that humans will face on the Moon coalesce nations to shape a blueprint together, a shared road map for expanding human presence throughout the solar system? Perhaps as we take steps to return to the Moon, we will seize the opportunity to ascertain common objectives for the betterment of science and technology, recognizing the basic humanity within our unquenchable thirst to press onward and outward.

To this point, once cislunar space and the Moon are commercial hubs of activity, to what degree does this action become a valuable strategy for a nation, whether putting in place settlements on the Moon, setting up lunar mining operations, and creating interplanetary way stations? What is the military role in protecting, perhaps defending, these off-worldly and international commodities?

How these Rorschach-like perceptions play out is anybody's guess. Unquestionably, new oceans of exploration and knowledge lie ahead. As it is said in the Chinese proverb, "The journey of a thousand miles begins with one step." In July 1969, Apollo 11's Neil Armstrong made that one small step . . . and there are many more to follow.

Adage aside, I do recall my late mother telling me, during the space race heyday of the 1960s, "The meek shall inherit the Earth . . . but the brave ones are going to the Moon!"

AFTERWORD

MISSION CONTROL

BY THE HONORABLE HARRISON H. SCHMITT

LUNAR MODULE PILOT, APOLLO 17, AND FORMER U.S. SENATOR

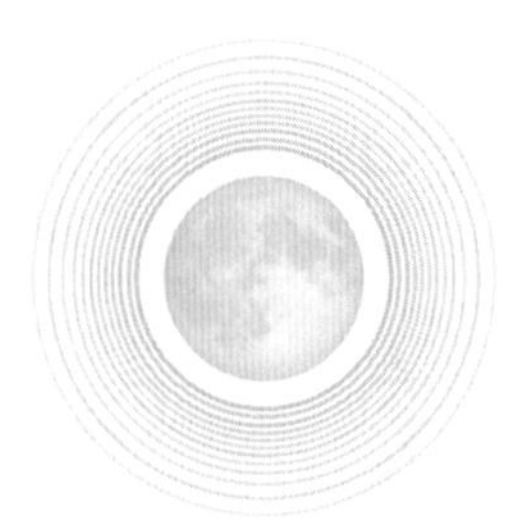

WHAT IS THE VALUE of returning to the Moon? That question forms the backbone of this book. As Leonard David has shown us, in addition to their immense geopolitical value, new lunar missions offer a range of scientific benefits, as well as a means of dealing with many of the challenges presented by our plans for the human exploration of Mars. The Moon lies only three days away, making it a convenient location for Mars mission development, simulation, and training. In addition, the importance of future Moon missions as the United States and its international and commercial partners explore, settle, and develop the new ocean of space cannot be overemphasized. Flying to the Moon and working there require the deep-space operational discipline and engineering risk management approach that made Apollo successful. New generations of space managers, engineers, and flight controllers will need to relearn and apply the lessons of humankind's first venture away from the gravitational confines of Earth as they prepare for future exploratory departures.

To move forward, a focused, Apollo-style management system will be needed. This system must involve people willing to, as the saying puts it, "Stay young, stay lean, stay risk takers." Once the decision to go back to the Moon and on to Mars is made, the sole focus should be to do just that.

What was the recipe for Apollo's success?

In those space race times, first of all, a geopolitical imperative required space leadership by America and the forces of freedom. NASA operated with a single-minded focus, made possible by a reservoir of young engineers and skilled workers, people working 16-hour days, seven-day weeks.

Another necessary part of that recipe was a base of technology with presidential support through four administrations, as well as congressional support for an adequate reserve of funding, roughly 100 percent of the total funding originally thought to be required to meet the stated goal of landing a man on the Moon.

Tough, competent, disciplined, and courageous management kept the program on track, even after the tragic on-the-pad Apollo 1 fire in January 1967, the aborted flight of Apollo 13 in 1970, and despite hardware development challenges, waning media interest, and late mission cancellations. An essential ingredient was minimally layered decision making. A good idea to correct a problem or enhance a mission could move rapidly to a decision, sometimes overnight.

The Apollo program drew upon a strong core of NASA engineering expertise, and strong, capable commercial contractor and supplier teams—although NASA controlled the development and manufacture of all items in the critical path to success. International participation was encouraged only for science experiments and lunar sample analysis.

The Apollo working environment reflected the principles of liberty. No one with whom I worked felt inhibited in speaking up at any time, from preliminary design reviews to critical design reviews to tests to simulations to flight operations.

All this emphasis on focus brought results. Between September 1968 and November 1969—Apollo 7 through Apollo 12—NASA, and its contractors and their suppliers, representing 450,000 young men and women mostly in their 20s, launched a Saturn V-class mission on the average of once every two months.

Fast-forward to today.

I am very pleased that President Trump and Vice President Pence have provided a Space Policy Directive that states: "Beginning with missions beyond low-Earth orbit, the United States will lead the return of humans to the Moon for long-term exploration and utilization, followed by human missions to Mars and other destinations."

Implementation of this policy is geopolitically critical and certainly as important to national security as the initiative that forced rapid implementation of Apollo in the 1960s. But because this new presidential initiative will reach into future generations, unlike the 10-year span of Apollo, several necessary revisions to the Apollo management environment are needed for success:

- The implementing agency must maintain an average age less than about 30. The military, such as the nuclear Navy, already does this, so modification of civil service rules for national security programs has a long precedent.
- A significant funding reserve, probably about 30 percent beyond implementation budgets, needs to be provided. The Office of Management and Budget (OMB) and Congress must advocate and protect these

reserves! They shield the milestones in schedules from unanticipated hardware and software challenges—the unknown unknowns that always appear.

- The private sector should maintain its traditional role in deep-space exploration, conducting independent study and serving as prime contractors and suppliers of computational and hardware capabilities. In these roles, contractors must meet the risk management requirements demanded by a national program.
- International participation should be encouraged, especially relative to science and engineering payloads and components and subsystems of low technical risk. No international hardware, software, or approvals, however, should be in the critical path to success.
- Commercial and international entities could take on a new role, shouldering the load of space activities that are not directly related to a geopolitical commitment to deep-space exploration. These new activities would include not only satellite communications but also data acquisition for space and terrestrial science, environmental monitoring, and space facility supply.
- Commercial and private entities also should take the lead in planetary pioneering, lunar and planetary base supply, space resource production, and settlement.

So now what needs to happen to guarantee implementation of the president's Space Policy Directive?

I have come to the conclusion that the president must either transfer all unrelated NASA programs and projects to other relevant agencies and re-create in NASA the management efficiency of the Apollo program or create a new National Space Exploration Administration modeled after the NASA that existed in 1969. Whichever course is chosen, it must be done soon and without hesitation.

This book has provided a look at the science and technology in play as humankind reconnects with the Moon—this time to stay, to survive and thrive there by utilizing lunar resources and the initiative and creative energy of human pioneers.

In this migration of the human species into space, it should be remembered that throughout history, plans that designate common access to natural resources have a notorious record of failure and ultimately do not result in full productivity. Systems that recognize private property have provided far more benefit to the world than those that attempt to provide common ownership.

Whenever and however a return to the Moon occurs, one thing is assured. That return will be historically comparable to the movement of our species out of Africa about 150,000 years ago. If led by an entity representing the democracies of the Earth, our return to the Moon will rejuvenate the foundations of self-government that continue to enhance human well-being.

APPENDIX

Lunar Exploration Time Line

Adapted from NASA Space Science Data Coordinated Archive (NSSDCA), NASA Goddard Space Flight Center, Greenbelt, Maryland.

1959

USSR Luna 1 - Jan 2, 1959 - Flyby
USA Pioneer 4 - Mar 3, 1959 - Flyby
USSR Luna 2 - Sep 12, 1959 - Impact
USSR Luna 3 - Oct 4, 1959 - Probe

1961

USA Ranger 1 - Aug 23, 1961 - Attempted Test Flight
USA Ranger 2 - Nov 18, 1961 - Attempted Test Flight

1962

USA Ranger 3 - Jan 26, 1962 - Attempted Impact
USA Ranger 4 - Apr 23, 1962 - Impact
USA Ranger 5 - Oct 18, 1962 - Attempted Impact

1963

USSR Luna 4 - Apr 2, 1963 - Flyby

1964

USA Ranger 6 - Jan 30, 1964 - Impact
USA Ranger 7 - Jul 28, 1964 - Impact

1965

USA Ranger 8 - Feb 17, 1965 - Impact

USA Ranger 9 - Mar 21, 1965 - Impact

USSR Luna 5 - May 9, 1965 - Impact

USSR Luna 6 - Jun 8, 1965 - Attempted Lander

USSR Zond 3 - Jul 18, 1965 - Flyby

USSR Luna 7 - Oct 4, 1965 - Impact

USSR Luna 8 - Dec 3, 1965—Impact

1966

USSR Luna 9 - Jan 31, 1966 - Lander

USSR Luna 10 - Mar 31, 1966 - Orbiter

USA Surveyor 1 - May 30, 1966 - Lander

USA Lunar Orbiter 1 - Aug 10, 1966 - Orbiter

USSR Luna 11 - Aug 24, 1966 - Orbiter

USA Surveyor 2 - Sep 20, 1966 - Attempted Lander

USSR Luna 12 - Oct 22, 1966 - Orbiter

USA Lunar Orbiter 2 - Nov 6, 1966 - Orbiter

USSR Luna 13 - Dec 21, 1966 - Lander

1967

USA Lunar Orbiter 3 - Feb 4, 1967 - Orbiter

USA Surveyor 3 - Apr 17, 1967 - Lander

USA Lunar Orbiter 4 - May 8, 1967 - Orbiter

USA Surveyor 4 - Jul 14, 1967 - Attempted Lander

USA Explorer 35 (IMP-E) - Jul 19, 1967 - Orbiter

USA Lunar Orbiter 5 - Aug 1, 1967 - Orbiter

USA Surveyor 5 - Sep 8, 1967 - Lander

USA Surveyor 6 - Nov 7, 1967 - Lander

1968

USA Surveyor 7 - Jan 7, 1968 - Lander
USSR Luna 14 - Apr 7, 1968 - Orbiter
USSR Zond 5 - Sep 15, 1968 - Return Probe
USSR Zond 6 - Nov 10, 1968 - Return Probe
USA Apollo 8 - Dec 21, 1968 - Crewed Orbiter

1969

USA Apollo 10 - May 18, 1969 - Orbiter
USSR Luna 15 - Jul 13, 1969 - Orbiter
USA Apollo 11 - Jul 16, 1969—First Crewed Landing
USSR Zond 7 - Aug 7, 1969 - Return Probe
USA Apollo 12 - Nov 14, 1969 - Crewed Landing

1970

USA Apollo 13 - Apr 11, 1970 - Crewed Landing (aborted)
USSR Luna 16 - Sep 12, 1970 - Sample Return
USSR Zond 8 - Oct 20, 1970 - Return Probe
USSR Luna 17 - Nov 10, 1970 - Rover

1971

USA Apollo 14 - Jan 31, 1971 - Crewed Landing
USA Apollo 15 - Jul 26, 1971 - Crewed Landing
USSR Luna 18 - Sep 2, 1971 - Impact
USSR Luna 19 - Sep 28, 1971 - Orbiter

1972

USSR Luna 20 - Feb 14, 1972 - Sample Return

USA Apollo 16 - Apr 16, 1972 - Crewed Landing
USA Apollo 17 - Dec 7, 1972—Last Crewed Landing

1973
USSR Luna 21 - Jan 8, 1973 - Rover
USA Explorer 49 (RAE-B) - Jun 10, 1973 - Orbiter

1974
USSR Luna 22 - Jun 2, 1974 - Orbiter
USSR Luna 23 - Oct 28, 1974 - Lander

1976
USSR Luna 24 - Aug 14, 1976 - Sample Return

1990
JAPAN Hiten - Jan 24, 1990 - Flyby and Orbiter

1994
USA Clementine - Jan 25, 1994 - Orbiter

1997
PEOPLE'S REPUBLIC OF CHINA/USA AsiaSat 3/HGS-1 - Dec 24, 1997 - Lunar Flyby

1998
USA Lunar Prospector - Jan 7, 1998 - Orbiter

2003
EUROPE SMART 1 - Sep 27, 2003 - Lunar Orbiter

2007

JAPAN Kaguya (SELENE) - Sep 14, 2007 - Lunar Orbiter

CHINA Chang'e 1 - Oct 24, 2007 - Lunar Orbiter

2008

INDIA Chandrayaan-1 - Oct 22, 2008 - Lunar Orbiter

2009

USA Lunar Reconnaissance Orbiter - Jun 18, 2009 - Lunar Orbiter

USA LCROSS - Jun 18, 2009 - Lunar Impactor

2010

CHINA Chang'e 2 - Oct 1, 2010 - Lunar Orbiter

2011

USA Gravity Recovery And Interior Laboratory (GRAIL) - Sep 10, 2011 - Lunar Orbiter

2013

USA Lunar Atmosphere and Dust Environment Explorer (LADEE) - Sep 6, 2013 - Lunar Orbiter

CHINA Chang'e 3 - Dec 1, 2013 - Lunar Lander and Rover

2014

CHINA Chang'e 5 Test Vehicle - Oct 23, 2014 - Lunar Flyby and Return

2018

CHINA Queqiao - May 20, 2018 - Lunar Relay Satellite

CHINA Chang'e 4 - Late 2018 - Lunar Far Side Lander

2019 on

SOUTH KOREA Korea Pathfinder Lunar Orbiter

CHINA Chang'e 5 Lunar Sample Return Mission

INDIA Chandrayaan-2 Moon Orbiter, Lander, and Rover

INDIA TeamIndus Lunar Lander (Private Mission)

ISRAEL SpaceIL Lunar Lander (Private Mission)

USA Astrobotic Peregrine Lunar Lander (Private Mission)

USA Moon Express (Private Mission)

RUSSIA Lunar 25 (Lander); Luna 26 (Orbiter); Luna 27 (Lander); Luna 28 (Lander/Sample Return); Luna 29 (Lander/Rover)

ACKNOWLEDGMENTS

Special thanks to David Kring, Lunar and Planetary Institute; James Head, Brown University; Ian Crawford, Birkbeck College London; Bernard Foing, European Space Agency; Angel Abbud-Madrid, Center for Space Resources at the Colorado School of Mines; Mark Robinson, Arizona State University; William Hartmann, Planetary Science Institute; Clive Neal, University of Notre Dame; and the solid footing of Apollo moonwalkers Buzz Aldrin and Harrison Schmitt for their assistance in writing this book.

I am deeply grateful for the editorial direction and support of Susan Tyler Hitchcock of National Geographic, as well as the eagle editing eyes of Liz Kruesi and Mary Norris. Thank you Susan Blair, Michelle Cassidy, Judith Klein, Heather McElwain, Ann Day, Daneen Goodwin, Katie Olsen of National Geographic for your tireless help. As always, my wife, Barbara David, was a constant light, brighter than the Moon itself, and made sure I didn't crater during this undertaking.

The following individuals provided welcomed guidance and expertise during the preparation of various drafts of this volume:

Chapter 1

Justin Arthur, The Walt Disney Archives; Ron Miller, space artist historian; Ray Williamson, archaeoastronomer; G. Jeffrey Taylor and Linda Martel of the Hawai'i Institute of Geophysics and Planetology, University of Hawai'i at Manoa

Chapter 2

David Blewett, Johns Hopkins University Applied Physics Laboratory; Anthony Colaprete, NASA Ames Research Center; Clark Chapman, Southwest Research Institute; David Lawrence, Johns Hopkins University Applied Physics Laboratory; Paul Lucey, Hawai'i Institute of Geophysics and Planetology, University of Hawai'i at Manoa; Jay Melosh, Purdue University

Chapter 3

James Burke, former JPL; Ron Creel, Apollo Lunar Roving Vehicle Team Member; Jay Gallentine, space historian; Dan Lester, Exinetics; David Portree, space historian; Jeffrey Richelson, National Security Archive

Chapter 4

John Cain, consultant, health risk management; Andrew Chaikin, space historian; Dean Eppler, The Aerospace Corporation; Ken Glover, Apollo Lunar Surface Journal; Michelle and Tim Hanlon, For All Moonkind; Eric Jones, Apollo Lunar Surface Journal; Ryan Kobrick, College of Aviation, Embry-Riddle Aeronautical University; Roger Launius, Launius Historical Services; Tom Murphy, University of California, San Diego; Robert Pearlman, collectSPACE; Irene Lia Schlacht, Politecnico di Milano, Italy

Chapter 5

Carlton Allen, NASA Johnson Space Center (Emeritus); Bradley Cheetham, Advanced Space; Leslie Gertsch, Missouri University of Science and Technology; Chris McKay, NASA Ames

Research Center; Philip Metzger, University of Central Florida, Florida Space Institute; Gerald (Jerry) Sanders, NASA Johnson Space Center; Alan Stern, Southwest Research Institute; Brian Wilcox, Jet Propulsion Laboratory; Ryan Zeigler, NASA Johnson Space Center

Chapter 6

Ben Bussey, NASA Headquarters; Jack Burns, University of Colorado, Boulder; James Carpenter, European Space Agency; Jason Crusan, NASA Headquarters; Todd Harrison, Center for Strategic and International Studies; Kaitlyn Johnson, Center for Strategic and International Studies; Kathy Laurini, NASA Johnson Space Center; John Logsdon, Space Policy Institute; Jamie Morin, Center for Strategic and International Studies; Brent Sherwood, Jet Propulsion Laboratory; Marcia Smith, Space PolicyOnline.com; Harley Thronson, NASA Goddard Space Flight Center; James Vedda, The Aerospace Corporation's Center for Space Policy and Strategy

Chapter 7

Charles Chafer, Celestis; A.C. Charania, Blue Origin; James E. Dunstan, Mobius Legal Group; Dan Hendrickson, Astrobotic Technology, Inc.; Alex Ignatiev, University of Houston; Michael Listner, Space Law & Policy Solutions; Jerome Pearson, STAR, Inc.; Virgiliu Pop, ESERO Romania, Romanian Space Agency

Chapter 8

Haym Benaroya, Rutgers University; Steve Durst, International Lunar Observatory Association; Peter Eckart, Advisory Board

Member, Buzz Aldrin Space Institute; Marc Gibson, NASA Glenn Research Center; Barbara Imhof, LIQUIFER Systems Group GmbH; Alexander MacDonald, Jet Propulsion Laboratory; Wendell Mendell, lunar savant; Julio Orta, artist; Jorge Mañes Rubio, artist; Paul van Susante, Michigan Technological University; Madhu Thangavelu, Graduate Space Concept Synthesis Studio, University of Southern California's School of Architecture; Dennis Wingo, Skycorp Incorporated

Note on pages 6–7: "The Original Moon," a poem by Carle Pieters, planetary scientist at Brown University, stems from a groundbreaking conference in 1984 in Kona, Hawaii, on the origin of the Moon—a meeting covered by *Science* reporter Dick Kerr who dubbed it the "Big Splash"—with Pieters inspired to pen the verses.

FURTHER READING

Aldrin, Buzz, and Ken Abraham, *Magnificent Desolation: The Long Journey Home From the Moon* (Three Rivers Press, 2010).

Amitai, Etzioni, *The Moon-Doggle* (Doubleday & Company, 1964).

Beattie, Donald, *Taking Science to the Moon: Lunar Experiments and the Apollo Program*, (Johns Hopkins University Press, 2001).

Chaikin, Andrew, *A Man on the Moon: The Voyages of the Apollo Astronauts* (Viking, 1994).

Cortright, Edgar M., ed., *Apollo Expeditions to the Moon* (Scientific and Technical Information Office, National Aeronautics and Space Administration, 1975).

Donovan, James, *Shoot for the Moon: The Space Race and the Extraordinary Voyage of Apollo 11* (Little, Brown and Company, 2018).

Dula, Arthur, and Zhang Zhenjun, eds., *Space Mineral Resources: A Global Assessment of the Challenges and Opportunities* (International Academy of Astronautics/Virginia Edition Publishing Company, 2015).

Eckart, Peter, ed., *The Lunar Base Handbook: An Introduction to Lunar Base Design, Development, and Operations,* 2nd ed. (McGraw-Hill Primis Custom Publishing, 2006).

Eicher, David J., and Brian May, *Mission Moon 3-D* (The MIT Press and the London Stereoscopic Company, 2018).

Harland, David M., *Exploring the Moon: The Apollo Expeditions* (Springer, 1999).

Heiken, Grant H., David T. Vaniman, and Bevan M. French, eds., *Lunar Sourcebook: A User's Guide to the Moon* (Cambridge University Press, 1991).

Jacobs, Robert, Michael Cabbage, Constance Moore, and Bertram Ulrich, eds., *Apollo: Through the Eyes of the Astronauts* (Abrams, 2009).

Kluger, Jeffrey, *Apollo 8: The Thrilling Story of the First Mission to the Moon* (Henry Holt and Company, 2017).

Logsdon, John M., *The Decision to Go to the Moon: Project Apollo and the National Interest* (The MIT Press, 1970).

Mendell, W.W., ed., *Lunar Bases and Space Activities of the 21st Century* (Lunar and Planetary Institute, 1985).

Muir-Harmony, Teasel, *Apollo to the Moon: A History in 50 Objects* (National Geographic, 2018).

Murray, Charles, and Catherine Bly Cox, *Apollo: The Race to the Moon* (Simon and Schuster, 1989).

Pop, Virgiliu, *Who Owns the Moon? Extraterrestrial Aspects of Land and Mineral Resources Ownership* (Springer, 2009).

Pyle, Rod, *Destination Moon: The Apollo Missions in the Astronauts' Own Words* (Collins/Smithsonian Books, 2005).

Reichl, Eugen, *Saturn V: America's Rocket to the Moon* (Schiffer Publishing, 2018).

Schmitt, Harrison, *Return to the Moon: Exploration, Enterprise, and Energy in the Human Settlement of Space* (Springer/Praxis Publishing, 2006).

Schrunk, David, Burton L. Sharpe, Bonnie L. Cooper, and Madhu Thangavelu, eds., *The Moon: Resources, Future Development and Settlement,* 2nd ed. (Springer/Praxis Publishing, 2008).

Scott, David Meerman, and Richard Jurek, *Marketing the Moon: The Selling of the Apollo Lunar Program* (The MIT Press, 2014).

Siddiqi, Asif A., *Challenge to Apollo: The Soviet Union and the Space Race, 1945–1974* (NASA History Division, 2000).

Spudis, Paul D., *The Value of the Moon: How to Explore, Live, and Prosper in Space Using the Moon's Resources* (Smithsonian Books, 2016).

Tumlinson, Rick N. ed., with Erin Medlicott, *Return to the Moon* (Collector's Guide Publishing/Apogee Books, 2005).

Wingo, Dennis, *Moonrush: Improving Life on Earth With the Moon's Resources* (Collector's Guide Publishing/Apogee Books, 2004).

Online Resources

Apollo Flight Journal *(https://history.nasa.gov/afj)*

Apollo Lunar Surface Journal *(www.hq.nasa.gov/alsj)*

European Space Agency *(https://www.esa.int/ESA)*

NASA History Office *(https://history.nasa.gov/)*

NASA Space Science Data Coordinated Archive *(https://nssdc.gsfc.nasa.gov/nmc/)*

Index

Index TK

INDEX

Index TK

Index TK

INDEX

Index TK

Index TK

INDEX

Index TK

Index TK

ABOUT THE AUTHOR

LEONARD DAVID is a space journalist who has reported on aerospace ideas, explorations, enterprises, and achievements for more than 50 years. Throughout those years, his writings have appeared in numerous website articles, newspapers, books, and magazines, including *Scientific American,* the *Financial Times, Foreign Policy, Private Air, Sky and Telescope, Astronomy,* and SPACE.com, as well as *Aerospace America,* and he has contributed supplemental writing for *Aviation Week & Space Technology* magazine. He has been a consultant to NASA, other government agencies, and the aerospace industry. In the mid-1980s, he served as director of research for the National Commission on Space.

David currently contributes to SPACE.com as the Space Insider columnist, serves as correspondent for *SpaceNews* magazine, and also maintains "Inside Outer Space" at *www.leonarddavid.com*. He is the author of *Mars: Our Future on the Red Planet* and co-author with Buzz Aldrin of *Mission to Mars: My Vision for Space Exploration.*

David lives with his wife, Barbara, in Golden, Colorado, where the clear, nighttime sky fuels the imagination about space travel to other worlds . . . as well as concern over lost luggage on the Moon or at Mars.

Back Ad TK